Microbiological Analysis of
Food and Water:
Guidelines for Quality Assurance

D1344495

J121637

Microbiological Analysis of Food and Water:
Guidelines for Quality Assurance

Edited by

N.F. Lightfoot
Public Health Laboratory Service
Newcastle upon Tyne
U.K.

E.A. Maier
European Commission
DG XII Measurement and Testing Programme
Brussels
Belgium

ELSEVIER
Amsterdam – Boston – London – New York – Oxford – Paris
San Diego – San Francisco – Singapore – Sydney – Tokyo

ELSEVIER SCIENCE B.V.
Sara Burgerhartstraat 25
P.O. Box 211, 1000 AE Amsterdam, The Netherlands

First edition (hardbound) 1998; Second impression 1999; Third impression 2003
Paperback edition 1999; Second impression 2003

Library of Congress Cataloging-in-Publication Data
Microbiological analysis of food and water : guidelines for quality
Assurance / edited by N.F. Lightfoot, E.A Maier.
    p.   cm.
    Includes bibliographical references and index.
    ISBN 0-444-82911-3 (hardcover : alk. Paper) -- ISBN 0-444-50203-3 (pbk. : alk. Paper)
    1. Food--Microbiology. 2. Water--Microbiology. 3. Food industry
and trade--Quality control. 4. Water quality. I. Lightfoot, N. F. (Nigel F.), 1945-
II. Maier, E. A.
    QR115.M46   1998
    664'.001'579--dc21
                                98-13035
                                CIP

ISBN: 0-444-82911-3 (hardcover)
ISBN: 0-444-50203-3 (paperback)

♾ The paper used in this publication meets the requirements of ANSI/NISO Z39.48-1992 (Permanence of Paper).
Printed in The Netherlands.

Preface

In 1986, the European Commission began to support research and development activities in the field of microbiological measurements. Within the BCR (Community Bureau of Reference), and later within the Measurement and Testing Programme (M & T Programme), several projects were launched which covered various aspects of food and water microbiology.

The main objectives of these projects were to develop adequate tools for the validation of microbiological measurement and testing methods. Two projects — food and water microbiology — included the development of representative and stable reference materials, and involved more than one hundred laboratories from the European Union in interlaboratory studies. Fundamental research on the behaviour of microbes in spray-dried milk powder was undertaken. Measurement methods and statistical treatment tools were developed to assess the homogeneity and stability of these reference materials, and the results led to the development of several reference materials used in European-wide interlaboratory studies. Several bacterial strains were studied in food (salmonella, listeria, *B. cereus*, *S. aureus*, *C. perfringens*, *E. coli*) and water (*E. faecium*, *E. cloacae*, *E. coli*, salmonella and *C. perfringens*). For each of the studied microbes, several methods of measurement or testing were compared. Quality assurance (QA) and quality control (QC) items have been developed. The major achievements of these projects were:

(i)   the production and validation of several standard operating procedures for basic components of microbiological testing or measurement methods (e.g., measurement of pH of culture media, measurement of temperature in incubators, etc):

(ii)   the development and validation of working tools for the proper
        organization of interlaboratory studies (transport of samples,
        analytical protocols, etc) and statistical treatment of the results;

(iii)   the production and certification of six reference materials.

In addition to these two projects, the Commission has also supported a study on the determination of faecal contamination of sea-
water. As the stability of microbes and the sea-water matrix could
not be guaranteed in a large-scale interlaboratory study (instability
during transport), the evaluation of the methods was organized in
one central laboratory where the participants applied their own
methods to common sea-water samples.

Finally, research projects on the development and validation of
new, rapid microbiological measurement methods were also supported by the Measurement and Testing Programme.

Within these projects, discussions revealed the necessity of developing several important procedures for proper quality assurance and
quality control systems in microbiological laboratories. Many of
these missing procedures were already known in other fields of measurement sciences, but had not been used in microbiology. Some QA
and QC principles and tools were already known, but were not readily available to all microbiologists. Finally, the knowledge developed
in the projects by the participating laboratories needed to be made
available to all testing laboratories through the development of a
complete set of guidelines. The importance of these guidelines was
reinforced by the implementation of national accreditation systems
and their mutual recognition within the European Union following
the resolution (90/C1/01 of 21 December 1989) of the European
Council on the 'global approach to conformity assessment'. The
accreditation systems are based on European Standard EN 45001,
which lists several requirements to be fulfilled by laboratories. Several procedures (e.g., statistical control, use of reference materials,
validation of methods, proficiency testing, etc.) are still not used by
microbiologists, and need to be translated into microbiological terms
or adapted to the particular situation of microbiological measurements or testing.

Some participants in previous BCR projects recognized the importance of the task. With the help of other leading QA and QC

microbiology specialists in Europe, they decided to prepare a complete set of guidelines on how to start and implement a quality system in a microbiological laboratory. This project has been supported by the Commission through the Measurement and Testing Programme.

The working group included food and water microbiologists from various testing laboratories, universities and industry, as well as statisticians and QA and QC specialists in chemistry.

This book represents the outcome of their work. It has been written with the express objective of using simple but accurate wording, to be accessible to all microbiology laboratory staff. To facilitate reading, the more specialized items, in particular some statistical treatments, have been added as an annex to the book. All QA and QC tools mentioned within these guidelines have been developed and applied by the authors in their own laboratories. All aspects dealing with reference materials and interlaboratory studies have been taken in a large part from the projects conducted within the BCR and M & T programmes of the European Commission.

The European Commission and the editors wish to express their sincere thanks to the authors of these guidelines for their tremendous effort. Thanks are also due to the participants and organizers of the former microbiology projects of the Commission, who produced the basic information and materials on which several aspects and conclusions of these guidelines are based.

It is appreciated that with so many different quality control procedures, their introduction in a laboratory would appear to be a formidable task. The authors recognize that each laboratory manager will choose the most appropriate procedures, depending on the type and size of the laboratory in question. Accreditation bodies will not expect the introduction of all measures, only those that are appropriate for a particular laboratory.

N.F. Lightfoot
Newcastle

E.A. Maier
Brussels

List of contributors

N.F. Lightfoot
Public Health Laboratory Service, Newcastle upon Tyne, U.K.
E.A. Maier
European Commission, DG XII Measurement and Testing
Programme, Brussels, Belgium
H. Beckers
Unilever Research Laboratory, Vlaardingen, Netherlands
S. Dahms
Institut fur Biometrie der freien Universität Berlin, Berlin, Germany
J.M. Delattre
Institut Pasteur de Lille, Lille, France
A. Havelaar
National Research Institute of Public Health & Environmental
Protection (RIVM), Bilthoven, Netherlands
M. Michels
Unilever Research Laboratory, Vlaardingen, Netherlands
S. Niemela
University of Helsinki, Helsinki, Finland
J. Papadakis
Athens School of Hygiene, Athens, Greece
D. Roberts
Food Hygiene Reference Laboratory, Central Public Health
Laboratory, London, U.K.
H. Tillett
Communicable Disease Surveillance Centre, Public Health
Laboratory Service Board, London, U.K.
H.R. Veenendaal
KIWA NV, Nieuwegein, Netherlands
H. Weiss
Institut fur Biometrie der freien Universität Berlin, Berlin,
Germany

Contents

Chapter 1

Scope and purpose

1.1. Introduction

Each day many laboratories throughout Europe will carry out thousands of microbiological analyses of food and water. Manufacturers' laboratories monitor the microbiological quality of raw materials, the effectiveness of treatment processes, critical control points during production and the final quality of the end product. Public laboratories monitor the quality of food, drinking water and bathing and recreational waters to determine whether existing national or international guidelines and standards are being satisfied. In addition, they may be involved in the investigation of customer complaints or in the investigation of suspected food-borne or water-borne disease. International standards for the microbiological quality of water and food are primarily those formulated by the EU and WHO. These standards set the criteria for the occurrence of pathogenic or potentially pathogenic bacteria in food and water but more commonly for the so-called indicator organisms or the total count of naturally occurring non-pathogenic bacteria.

1.2. Implications of incorrect results

This vast number of microbiological analyses indicates the importance of quality assurance programmes in the laboratories carrying out these tests. Incorrect test results may have extensive economic or public health impact. False-positive results may lead to unnecessary condemnation of batches of food products, the unnecessary closing of recreational water facilities or unnecessary corrective actions in drinking-water supply such as widespread notices advising the boiling of water. These measures may be very costly. The impact on the public will be to arouse unnecessary concerns and worries; the impact on the producer/manufacturer will be unnecessary damage to their reputation and consequent loss of products.

False-negative results may expose the public to direct risks to public health due to the release of contaminated food products, the consumption of inadequately treated water or swimming in polluted water. In addition, false negative or consistently inaccurate low results may lead to improper competitive positions.

It is difficult to estimate accurately the real impact of measurements and of their quality in all fields of economy and social activities. Some authors have made an attempt to estimate figures based on the importance of clinical measurements in industry or clinical chemistry (Uriano and Gravatt, 1977). These estimates can be extended to microbiological determinations.

H.S. Hertz (1988) has estimated that quality of analysis directly affects between 0.5 and 8% of the gross national product, and that about 10% of measurements are of poor quality and have to be repeated; in 1988, Hertz estimated these repetitions to cost $5 billion per year in the United States. G. Tölg (1982) evaluated that Germany spent DM 12 billion in 1982 for repeated measurements. All these figures do not take into account the economical and social effects of bad measurements. These cannot be evaluated easily in terms of budget.

1.3. Quality assurance

The purpose of these guidelines is to facilitate the implementation of quality assurance programmes in all laboratories involved in food, water and environmental microbiology. This can be achieved only by a thorough understanding of the whole process and the factors that will affect the various constituent parts of the sampling and measurement methods. These are summarized in the process diagram in Fig. 1.

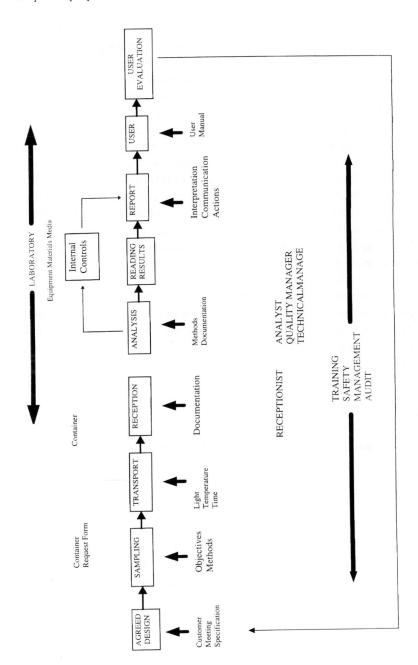

Fig. 1. Quality assurance programmes — processes to be considered.

These guidelines will present in each chapter an expanded description of each of these steps to help laboratory managers and quality assurance managers understand where errors can arise. Appropriate steps can then be taken to control these sources of error and a quality assurance programme can be written.

Quality assurance is defined by the International Organization for Standardization (ISO) as "all those planned and systematic actions necessary to provide adequate confidence that a product, process or service will satisfy given quality requirements". It includes quality control — "the operational techniques and activities that are used to fulfil requirements for quality". The operational techniques applied to the analytical method are referred to as analytical quality control (AQC) and include the use of negative controls (blanks), positive controls and reference materials (RMs) of known levels of contamination. External quality assessment (EQA) schemes provide the same samples of unknown content to a large number of laboratories so that performance can be compared. All these mechanisms to improve quality are at the early stages of development and quality assurance of the overall process is only just beginning. These guidelines will allow each laboratory to prioritize the steps necessary to improve quality and begin their implementation.

1.4. Types of laboratories

Laboratories carrying out these investigations will vary in size and complexity depending on the type and size of workload. Some may be large laboratories carrying out many process and final product checks in a large food manufacturing plant or in a large water company, others may be small laboratories with just two or three members of staff, or even be part of chemistry laboratories with one member of staff. It may seem daunting, therefore, to implement a quality assurance programme in these smaller laboratories; however, in the larger laboratories the QA programme will be extensive whereas in the smaller laboratories it will be quite manageable. The value of always having confidence in the results produced by a laboratory is probably cost-effective whatever the size of the laboratory when one considers the implications of a wrong result and the consequent management that may be required in such a situation. The important step for all laboratories is to understand the measurement process in its entirety, break it down into steps, prioritize the steps and use these guidelines to design the elements necessary for a quality assurance programme.

1.5. Elements of a quality assurance programme

The implementation of a quality assurance programme, perhaps the most important part of this book, is the subject of Chapter 2. This chapter outlines a method for dissecting the various work processes and to obtain an understanding of all factors that may affect the reliability of the process. There is advice on prioritization and a starting point for the programme implementation. Examples of documentation are given.

The concepts involved in the methods used in food and water microbiology are described in Chapter 7. The methods, their objectives and limitations as well as definitions of accuracy, trueness and precision are given and explained. The possible sources of error will become apparent and AQC involving first-, second- and third-line checks are described in Chapter 8.

The implementation of these checks involving the laboratory staff, their organization, responsibilities and management is the subject of Chapter 3. In addition, the general principles of laboratory design are explained and the importance of safety stressed. In any quality assurance programme the accountability of the staff to each other and to management, in a clear organizational structure, is important. The staff should be adequate in number and sufficiently trained to undertake the particular workload of a laboratory, but, just as importantly, staff should be motivated.

The quality of the sample cannot be underestimated. In Chapter 4, the whole process of sampling is broken down into its constituent parts so that potential deleterious effects can be identified and controlled. It is important to understand the reasons for and the objectives of the sampling so that a sampling plan can be agreed between the client and the laboratory. If a sample is of poor quality or mishandled on its way to the laboratory, the analysis is compromised. In some situations, the samples are taken outside the control of the laboratory and delivered with an unknown history. This practice is not satisfactory in a quality programme and steps must be taken to introduce quality and checks into the sampling process. In many standards, temperature controls during transport are prescribed and temperature checks will be necessary at sampling and on receipt in the laboratory. There may also be limits for sample transport times.

The reliability of the performance of equipment used in the microbiology laboratory is considered in Chapter 5. Items of equipment are used throughout the whole process, from the sterilization of materials and preparation of media to filtration of samples and incubation of cultures, and in some situations to the counting of colonies. The analyst plays a particularly important role in the reading of many of the results and relies on his/her skill and training. This can be nullified by the effects of equipment that is not controlled. The specifications of the equipment are given and the control actions necessary for quality are described.

The materials used, their specification and testing are the subject of Chapter 6. The descriptions provide a valuable insight into the possible deleterious effects on results that these materials may have, even though they may often be considered unimportant in quality assurance. They should be understood in the context of the whole process and checks included in the quality assurance programme.

Quality control of chemical analysis relies at several levels on statistical tests, e.g. calibration evaluation with single or multiple standards, various types of control charts with reference materials, duplicate analyses, use of certified reference materials to assess the accuracy of a method or proficiency testing. The application of statistics in quantitative microbiology is in its infancy and Chapter 8 gives comprehensive details of duplicate counting, parallel plating, dilution series and split samples. The use of reference materials and interlaboratory comparisons using external quality assessment schemes are described and the annexes to Chapter 8 give details of the statistical methods that can be used, and worked examples are demonstrated. Control charts can then be introduced and, though their use is still being evaluated, it is hoped that they will demonstrate when problems with the control of the analyses begin to occur.

Obtaining a result which should be correct is not the end of the quality assurance process. The handling and reporting of results also needs particular attention. Chapter 9 deals with the actions that should be taken once a result has become available, how it should be recorded and how it should be reported. Interpretation of the result will often be required, for follow-up actions may be necessary.

1.6. Accreditation

All laboratories that carry out microbiological testing in support of enforcement of regulations must eventually be accredited by the appropriate accreditation body. The requirements that are applied to these laboratories are given in the EN 45000 series of standards. A Council Resolution of the European Union on 'A global approach to conformity assessment' (1990) recommends that Member States set up accreditation systems based on the EN 45000 (EN 45001 for testing,) series of standards (1989). This resolution resulted in the recognition agreements of the accreditation systems between Member States to help a better implementation of the European internal market. The description of these standards and the addresses of the different national bodies are given in Chapter 11. The impact of a first reading of these documents is to fill laboratory personnel with dismay, giving them what at first sight appears to be an insurmountable task. Once the decision is taken to prepare for accreditation, the task soon becomes much simpler and the preparative work is the major benefit of accreditation. The reader of these guidelines who introduces a quality assurance programme will find the accreditation task much simpler. These guidelines, however, are not mandatory and should not form in their totality a basis for accrediting; they seek to explain the whole process and allow the laboratory to choose which control measures are appropriate for the type of work that it performs.

1.7. Benefits of a quality assurance programme

The key to a quality assurance programme is a thorough under-standing of the total process of a particular examination, from sampling to reporting of results, so that appropriate quality measures can be applied to the most critical points. These guidelines will provide the means to make that assessment. The choice of quality procedures to be implemented must be made by the quality manager in consultation with the laboratory staff. Prioritization will be important. It is impossible to introduce every measure on the first day but there is no doubt that the process will be revealing and lead to motivated staff with a good understanding of their importance in the function of the laboratory. It will lead to confidence in results and this is particularly important when unexpected results occur. It will satisfy staff and hopefully lead to valid comparisons across Europe. Once a quality assurance programme has been implemented and its benefits begin to become apparent, the efforts that were required in the early stages will be forgotten.

References

CEN, 1989. General criteria for the operation of testing laboratories (European Standard 45001), CEN/CENELEC, Brussels, Belgium.

Council Resolution of 21 December 1989 on a global approach to conformity assessment (90/C 10/01), Official Journal No. C 10/1, 16 January 1990.

Hertz, H.S., 1988. Are quality and productivity compatible in the analytical laboratory. Anal. Chem., 60(2): 75A–80A.

Tölg, G., 1982. ISAS Dortmund, private communication.

Uriano, G.A. and Gravatt, C.C., 1977. The role of reference materials in chemical analysis. CRC critical review. Anal. Chem., Oct. 1977: 361–411.

Chapter 2

Implementation of quality assurance programmes

2.1. Introduction

The implementation of a quality assurance programme starts with the nomination of a coordinator, who will be in charge of producing the quality assurance manual. The coordinator will lead the implementation programme and, in effect, becomes the quality manager. The relation between coordinator and laboratory management should be clearly defined and responsibilities agreed from the start. The success of the operation will depend to a large extent on the commitment shown by the laboratory management, as the setting up of a good QA programme will take the laboratory staff a significant amount of time.

The first action is to make an inventory of the status quo, i.e. a detailed statement of the existing situation in the laboratory, using the various chapters of this book as a reference.

Secondly, the analytical task of the laboratory (mission statement) should be well defined and agreed with the laboratory management to ensure that the objectives are well understood.

When there is a good understanding of the current practices and objectives, the implementation starts by assuring that all elements of a well-controlled QA programme are put in place and that weak or missing elements are fortified and built into the QA system.

The elements to be covered are given below in a logical order, but the sequence should be adapted to the local situation and another priority order may be established.

A time plan for the different tasks to be performed, with milestones to be met, will help to control the progress of the operation. Depending on the situation found, the implementation may take from a few months to one or two years.

It is recommended that for each task a small team of individuals is chosen and briefed by the coordinator to proceed with that particular task. This prevents the new procedures coming from the top, without the input of the experienced personnel who are responsible for the good execution of the daily operations. The format of all documents to be written should be agreed before the start of the operation.

2.2. Personnel (see Chapter 3)

An up-to-date organization diagram of the department forms the basis for the organization of the QA function. When that is drawn up or updated, a complete set of job descriptions for the individuals in the department should be made so that the responsibilities and authorities for the various tasks of the department can be clarified.

Once the tasks are known, it should be checked whether the skill base of the personnel in the department is sufficient for the jobs to be performed. This will lead to a review of the training needs of the personnel and, as training will take time, it is best to start with a training plan early on.

Once the task of the department has been defined, it may prove necessary to review the laboratory accommodation and facilities, to see whether these match the requirements resulting from this definition.

An important aspect to be checked is the safety requirements for the jobs performed. Testing for pathogens may require specific conditions, to be met as required by national or EU legislation. The appointment of a safety officer with the responsibility to supervise safety aspects of the laboratory is recommended.

2.3. Media (see Chapter 6)

The next step is to bring the quality assurance of the prepared media up to standard.

Media composition should be documented by detailed media recipes. When using dehydrated media or media components of commercial origin, the brands and codes should be detailed in the recipe.

To ensure the quality of dehydrated media and media components, manufacturers will perform their own quality control checks. Results of these quality control checks are summarized in certificates. Laboratories should make sure that batches of dehydrated media or media components are supplied with the relevant analysis certificates, as this is not a routine service. Ordering larger volumes of media of one batch number helps the control of the quality with minimal effort.

The procedure for the preparation of media should be precisely described and the sterilization thereof validated.

Routine checks on the quality of prepared media are sometimes thought to be superfluous as long as the relevant batches of dehydrated media have been checked (in house) before release and the preparation has been under control. The check before release of a batch may be carried out by the ecometric method (Mossel *et al.*, 1980) or by any other reliable quality control procedure (IUMS, 1982). A balance is required between process controls and end product testing.

A routine check on the pH of prepared media after sterilization is essential, but this in itself could be sufficient, assuming that the sterilization process has been under control.

When preparing media, batch numbers of commercial dehydrated media and/or components should be registered, as well as their date of preparation and final pH after sterilization. The sterilization process data should be stored linked with the sterilized batches of prepared media.

Finally, the date of final usage on the relevant batches of prepared media, based on the date of preparation and the shelf life of the relevant media, should be noted.

Shelf life and storage conditions should be part of the media preparation documentation.

2.4. Methods (see Chapter 7)

All methods should be available as written instructions in a uniform format and in the form in which they are used. Where standard
methods are followed, reference should be made to these standards. As
official standard methods tend to be lengthy, an abbreviated description for use at the bench could be drawn up, this summary giving the
key points of the method in the form of a flow sheet supported by a
short explanatory text; examples are given in the FAO Manual
(Andrews, 1992).

Interpretation and reading of plates may need extra attention.
The rules followed by the department should be made explicit and,
where confirmation is required, the local practice followed should be
specified. Where deviations from official standard methods have
become normal practice (as with no routine confirmation of *Entero-
bacteriaceae* from violet red bile glucose agar plates), this should be
evident from the documentation.

Minimum performance criteria should be established to allow
introduction of new methods in a controlled way.

2.5. Equipment (see Chapter 5)

All procedures should be in place for equipment calibration and
use; any checks that have to be done before their use in the laboratory should be specified.

All equipment should be registered; the frequency of maintenance
should be indicated here, as well as what this maintenance comprises
and who is responsible for carrying it out.

Equipment that has broken down should be labelled as such; its
repair should be recorded in the relevant equipment file.

Critical equipment, i.e. equipment that may influence, directly or
indirectly, the analysis results when not operating properly, should
be calibrated on a regular basis. Examples of critical equipment are
balances, pH meters, volume dosing equipment, thermometers, incubators, water baths and refrigerators. The frequency of calibration
should be documented as well.

The information relating to calibration, maintenance, etc. of the equipment may be recorded in instrument books for all the types of equipment available.

2.6. Sample handling (see Chapter 4)

Sampling starts with a good sampling instruction; development of a suitable protocol is a key activity. Furthermore, all requirements for sample transport and sample processing should be defined and implemented.

When sampling procedures are recorded, it may prove useful first to audit the actual sampling process to check that no unrealistic assumptions are made.

Where sampling is outside the control of the department, rules should be formulated about the processing of samples with an unknown sample history. It may not always be possible or necessary to reject such samples, but in the laboratory report this should be stated to prevent unjustified conclusions being drawn from the analytical findings.

2.7. Data handling (see Chapter 9)

All procedures for reading, recording and processing of microbiological data should be drawn up. The introduction of one or more uniform data collection sheets, in which the raw data are collected, should get priority. Once such a form is completed and signed by the analyst, the results can be transferred to a computer file or report form and reported to the requesting customer.

Preferably, data should not be communicated as analytical data alone; some interpretation of the findings should be added. The authority to interpret data should be clearly defined and interpretation should be restricted to data for which there is a good understanding of the sampling procedures and the actual product under investigation. Otherwise, the conclusions should be restricted to the sample analysed, without widening these to the material from which the sample was taken.

Before the interpretation of sample data, the results of any QC checks carried out with these samples should be taken into consideration. A system to ensure this should be set up as part of the laboratory organization.

Where statistical checks on the performance of the laboratory are set up, suitable means should be available to perform such tests and to check their actual significance on a regular basis, as more experience with such tests becomes available.

2.8.   Quality control systems (see Chapter 8)

Where quality control systems are to be established, three lines (or levels) should be distinguished. In the first line, the check should be carried out as a means of self-control for the analyst carrying out the analytical work and comprises, among others:
  – checks on the quality of prepared media;
  – calibration of critical apparatus;
  – use of control samples (positive, negative);
  – use of reference samples.
In the second line, the checks should enable evaluation of the repeatability and of the intra-laboratory reproducibility and comprises, among others, split samples or cultures with known characteristics.

In the third line, the checks are a part of an external quality assessment scheme. This 'proficiency testing' is the conclusive step of the quality assurance programme.

At present, proficiency testing is hampered by a lack of proper reference samples, but developments are ongoing. Reference samples are or will become available for proficiency testing on a daily basis.

For example, laboratories routinely examining samples for salmonella may use the BCR salmonella reference sample (certified or non-certified) day by day and interpret the results obtained over a period of time (see Chapter 8).

In the example given (Fig. 2.1, pp. 16 and 17), proper performance is demonstrated by two negative results in a four-week period. The calendar shows problems in the period of June and July. Once problems have been encountered, a search for their origin should be initiated. Apparently, the problems were solved in August.

Similar external reference samples will become available for other micro-organisms in the near future, not only for presence–absence tests, but also for enumeration procedures. For more information on these reference samples, refer to SVM details in Chapter 8 (Section 2.5).

Sometimes a shelf-stable sample is available of sufficient quantity and with sufficient contamination to serve as an internal reference sample.

With reference samples, the performance of the laboratory and its individual technicians may be checked on a regular basis. Taking into consideration repeatability/reproducibility data of the different types of analyses, Shewhart charts or other relevant systems to enable continuous evaluation of the performance of the laboratory may be developed.

An additional way of checking performance is participation in external quality assessment schemes or proficiency ring tests. Samples are prepared by an organizer and distributed; the requested analyses are carried out by the participants and the results are reported to the organizer. The organizer evaluates all incoming data and reports a summary to the participants.

As stated before, the organization of this type of proficiency ring test is hampered by the enormous workload to prepare proper reference samples. On the other hand, possibilities of participating in these proficiency ring tests are increasing, since ring tests have been recommended for laboratory accreditation. Even without considering laboratory accreditation, participation in these ring tests is very useful. They will develop and become more practicable and fit for different types of laboratories. External quality assessment schemes for food and water now exist in the United Kingdom and in the Nordic group of countries, and in France (water only).

2.9. Follow-up

When the previous steps are completed, all the material for a laboratory QA manual is prepared. All documentation should be produced in the same format and then passed through an internal acceptance procedure, so that the manual can be approved and its use implemented.

This completes the introduction of a QA/QC system; a first internal audit or validation can be arranged shortly thereafter. A formal

January

	53	1	2	3	4
MO		4□	11□	18□	25□
TU		5■	12□	19□	26□
WE		6□	13□	20□	27□
TH		7□	14□	21□	28□
FR	1	8□	15□	22□	29□
SA	2	9	16	23	30
SU	3	10	17	24	31

February

	5	6	7	8
MO	1■	8□	15□	22□
TU	2□	9□	16□	23□
WE	3■	10□	17□	24□
TH	4□	11□	18□	25□
FR	5□	12□	19□	26□
SA	6	13	20	27
SU	7	14	21	28

March

	9	10	11	12	13
MO	1□	8□	15□	22□	29■
TU	2□	9□	16□	23□	30□
WE	3□	10□	17□	24□	31□
TH	4□	11■	18□	25□	
FR	5□	12□	19□	26□	
SA	6	13	20	27	
SU	7	14	21	28	

April

	13	14	15	16	17
MO		5□	12	19□	26□
TU		6□	13□	20□	27□
WE		7□	14□	21■	28□
TH	1	8□	15□	22□	29□
FR	2□	9□	16□	23□	30□
SA	3	10	17	24	
SU	4	11	18	25	

May

	17	18	19	20	21	22
MO		3□	10□	17□	24□	31
TU		4□	11□	18□	25□	
WE		5■	12□	19□	26□	
TH		6□	13□	20□	27□	
FR		7□	14□	21□	28□	
SA	1	8	15	22	29	
SU	2	9	16	23	30	

June

	22	23	24	25	26
MO		7■	14■	21■	28■
TU	1□	8□	15□	22□	29■
WE	2■	9□	16□	23■	30□
TH	3□	10□	17□	24■	
FR	4□	11□	18□	25□	
SA	5	12	19	26	
SU	6	13	20	27	

Fig. 2.1. Presence absence testing for salmonella — daily results using BCR salmonella reference material. (□) =positive; (■) = negative.

July

		26	27	28	29	30
MO			5■	12■	19□	26□
TU			6□	13□	20□	27■
WE			7□	14■	21□	28□
TH		1■	8□	15□	22□	29□
FR		2□	9■	16□	23■	30□
SA		3	10	17	24	31
SU		4	11	18	25	

August

		30	31	32	33	34	35
MO			2□	9■	16□	23□	30□
TU			3□	10□	17□	24□	31□
WE			4□	11□	18□	25□	
TH			5□	12□	19□	26□	
FR			6□	13□	20■	27□	
SA			7	14	21	28	
SU		1	8	15	22	29	

September

		35	36	37	38	39
MO			6□	13□	20□	27□
TU			7□	14■	21□	28□
WE		1□	8□	15□	22□	29□
TH		2□	9■	16□	23□	30□
FR		3□	10□	17□	24□	
SA		4	11	18	25	
SU		5	12	19	26	

October

		39	40	41	42	43
MO			4□	11□	18□	25□
TU			5□	12□	19□	26□
WE			6□	13□	20■	27□
TH			7□	14□	21□	28□
FR		1□	8□	15□	22□	29□
SA		2	9	16	23	30
SU		3	10	17	24	31

November

		44	45	46	47	48
MO		1□	8□	15□	22□	29□
TU		2□	9□	16□	23□	30□
WE		3□	10□	17□	24■	
TH		4□	11■	18□	25□	
FR		5□	12■	19□	26□	
SA		6	13	20	27	
SU		7	14	21	28	

December

		48	49	50	51	52
MO			6□	13■	20□	27■
TU			7□	14□	21□	28□
WE		1□	8□	15□	22□	29□
TH		2□	9□	16□	23□	30□
FR		3□	10□	17□	24□	31□
SA		4	11	18	25	
SU		5	12	19	26	

Fig. 2.1. Continued.

updating procedure for the manual then needs to be agreed and implemented.

At this stage, the system itself will not yet be perfect and, depending on the objectives of the laboratory, the procedures and practices will have to be reviewed and/or improved every two years.

References

Andrews, W., 1992. FAO Manuals of Food Quality Control, 4 (Rev. 1), Microbiological Analysis. Food and Agricultural Organization of the United Nations, Rome.

IUMS (International Union of Microbiological Societies) and ICFMH (International Committee on Food Microbiology and Hygiene), 1982. Quality Assurance and Quality Control of Microbiological Culture Media. J.E.L. Corry (Ed.), G-I-T Verlag Ernst Giebeler, Darmstadt.

Mossel, D.A.A., van Rossem, F., Koopmans, M., Hendriks, M., Verouden, M. and Eelderink, I.. 1980. A comparison of the classical and the so-called ecometric technique for the quality control of solid selective culture media. J. Appl. Bacteriol., 49: 405–419.

Chapter 3

People, management and organization

3.1. Introduction

The main objective of a laboratory testing food or water is to produce reliable results on the samples examined. Assurance of the quality of those results is a prime responsibility of the manager of the laboratory and involves a wide range of tasks including:
- setting of objectives, both short- and long-term;
- establishment of measures to ensure the quality of results ;
- organization, management and motivation of people;
- supervision of technical performance;
- evaluation of the success and failure of individuals and procedures;
- resolution of complaints and anomalies, customer relations.

The achievement of a quality system involves encompassing all elements of quality control and quality assurance. The 'quality system' has been defined as the organization, structure, responsibilities, activities, resources and events that together provide organized procedures and methods of implementation to ensure the capability of the organization to meet quality requirements (Task Force Report, 1984). The quality achieved will be dependent to a great degree on the training, education and skill of the people involved in the analysis of samples and production of reports. This chapter will concentrate, therefore, on the essential points in relation to laboratory management and staffing necessary to ensure quality performance.

3.2. People

3.2.1. Management

All laboratories should have an organizational chart showing all members of staff, their roles in day-to-day work, the lines of management and responsibility and the relationship with other units or laboratory sections. It is important that all staff are aware of the extent and limitations of their responsibilities and also of their authority, and to whom they can delegate authority. Each staff member should have a well-defined job description that includes these limits of responsibility and authority and also the authority to evaluate the different levels of quality control checks described in Chapter 2. Detailed specifications of the qualifications, training and experience necessary for the job should be available for key managerial and technical staff and for all other staff where the absence of such documentation could impair the quality of tests. Job descriptions should be discussed, agreed, documented and signed for.

The workload of individual staff should be assessed in order to prevent overwork which can lead to errors and accidents. The availability of sufficient staff to undertake the work should be borne in mind when accepting samples for examination.

3.2.2. Qualifications of staff

The effective operation of the laboratory requires a range of different tasks which will require staff of different qualifications and training. It is useful to break down the steps in the examination of samples into clusters of tests so that the level of knowledge or training can be determined for working in these clusters. Training schemes will differ between countries and there are some tasks which may be carried out without a formal laboratory qualification but following training on-the-job.

In a food or water microbiology laboratory, there are two general types of technical personnel: analysts (technicians/scientists) who perform the actual microbiological examination of the sample, and support personnel who, with adequate training and supervision by analysts, prepare media and solutions, provide clean and/or sterile glassware and instruments, weigh test portions for examination and look after general laboratory hygiene (provision of disinfectant solutions, removal of contaminated tubes and plates, etc.). The latter group of staff should be trained by its supervisor and its duties well defined. Such non-qualified laboratory personnel must understand the tasks allocated, the importance of performing the duties at the necessary level to produce quality results and the necessity to report unexpected observations that may indicate situations which require the attention of a technically qualified member of staff.

The following is a suggested scheme for the grouping of tasks in relation to the training and education of laboratory personnel:

(i)   Preparatory work in relation to cleaning and disinfection of equipment and glassware, media preparation and distribution, laboratory hygiene, sample preparation.

Support personnel — no formal qualifications, ideally given training on an extended training course, or in-house training on-the-job.

Duties of suitably trained and experienced staff with knowledge of safety implications can be extended to simple tasks involving the handling of live organisms such as inoculation of tubes and plates.

(ii) Routine examinations according to standardized instructions —basic technical qualifications or undergoing training to achieve the basic qualification. Reading and recording of results by trainees should be supervised by a suitably qualified member of staff.

(iii) More complex examinations, independent work, supervision of trainees.

Analysts — full technical qualifications.

(iv) Interpretation of results: full technical qualifications with a defined period (e.g. three years) of practical experience of food/water examination. As different countries vary in their methods and requirements for technical qualifications this term has been used here in a very broad sense. It may cover a university degree in an appropriate subject such as microbiology, combined with some postgraduate experience/training, or a technical qualification obtained by attending courses on a day-release or block-release basis, or full-time courses with or without an industrial release component (sandwich course) at more technical institutions.

It is important that the laboratory defines which members of staff are trained and authorized to perform particular types of test and to operate particular types of equipment. An up-to-date record should be maintained of relevant competence, academic and professional qualifications, training and experience of all staff concerned with microbiological examinations. Training and competence should be checked by the designated member of staff responsible for the day-to-day management of the laboratory at the completion of the training period.

To implement and maintain a quality system, specific requirements for laboratory organization should include:

(i)   a designated member of staff with overall responsibility for the technical operation of the laboratory and for ensuring that the requirement of the quality system are met; and

(ii)   a designated member of staff with responsibility for ensuring the requirements of the quality system are met on a day-to-day basis and who will have direct access to the highest level of management at which decisions are taken on laboratory policy or resources, and also to the person responsible for the technical operation of the laboratory.

The laboratory organization should also have designated signatories, persons competent to sign the laboratory's test reports and to take technical responsibility for their content. Samples should not be accepted for examination or tests performed if the nominated personnel are not readily accessible to provide the necessary authorization and control and to take any critical decisions that may be required. There should also be a defined plan for provision of cover for work when staff members are absent.

3.2.3. *Training*

The training requirements for different levels of tests have already been outlined. It is important that new employees receive some form of orientation or induction at the commencement of employment. This will vary according to the previous training and experience of the particular staff member but must include sufficient technical detail to provide a basic understanding of the role of the laboratory, the basic elements of the job itself and the importance of the laboratory quality assurance programme. This is best achieved on a one-to-one basis with close monitoring by a supervisor. On-the-job training, where there is direct working with a colleague who knows the job well, is essential. Training in a particular skill or test ends when reproducible results are routinely obtained. The latter can be monitored by instituting the first- and second-line checks described in Chapter 8. The first-line checks are on-going self-control checks by the analyst. They should be carried out with every series of analyses (parallel plating, procedural blanks, positive and negative controls, etc.) and should be supervised by the direct superior of the analyst. For evaluating the success of training second-line checks, periodic checks by a person independent of the analyst can be introduced. The main methods are the introduction, unknown to the analyst, of a sample or culture of known characteristics or the use of split samples (replicate examination), examination of a further sub-sample from a product/material examined previously or examination of standard or reference samples.

On-the-job training develops skills and introduces local procedures. Assigning an individual to a position of greater responsibility within the laboratory organization may require further instruction either in house or by attending courses at recognized institutions. Whichever method is applicable, an up-to-date record must be kept of each training step and each qualification gained. Training records should list all the analyses and work carried out, and should be marked off when the individual is competent at each one. Job descriptions and responsibilities should be revised regularly to take into account these training achievements and qualifications.

Assessment of training and competence has hitherto been qualitative. There are now means to measure these by repeat examination/

parallel testing, use of control samples or reference materials. The tests described in Chapter 8 can be used for this purpose. Thus a level of achievement will be set that will give confidence to personnel. Plotting results on control charts will demonstrate the ability of the worker to maintain performance within control limits.

Where possible, staff should rotate between different sections of a laboratory so that they become familiar with all the procedures appropriate to their level of skill and training. Cover can then be provided in the absence of colleagues or where there is an unexpected increase in workload.

A designated training officer should be included in the laboratory organization. Attendance at scientific meetings, seminars and workshops offers opportunities to learn of new methods and techniques and can be recorded as part of the individual's continuing professional education. The continuing appraisal of staff performance is an essential element of a quality assurance system, yet it is often neglected or poorly performed. The primary objective is to improve job performance, but appraisal can also serve to motivate staff and assist supervisors in making decisions on training and development needs and staffing. Appraisal should be continuous but a formal appraisal, once or twice a year, is important. A formal appraisal may cover performance objectives, quality and productivity requirements, work safety, priority job elements, classification of responsibilities, personal long-term goals and barriers to good performance, and their causes and cures. The immediate goal is to agree upon a plan to help the particular staff members to improve their effectiveness.

3.3. Laboratory accommodation and environment

Production of quality results requires facilities that are at least adequate. The suitability of a laboratory, its location, design and environmental conditions, both internal and external, can influence the response of staff, the operation of sensitive equipment and ultimately the efficiency and effectiveness of a quality assurance programme.

It is not proposed to describe in this chapter the specific details of a microbiology laboratory, but only to mention the points which should be considered when assessing whether the laboratory and its environment are adequate for the production of quality results. The basic requirements should include:

(i)   sufficient space so that the work area can be kept clean and in good order;

(ii)   layout devised with efficiency in mind;

(iii) sufficient space for support facilities;

(iv)   office space for clerical staff;

(v)   cloakroom facilities for all staff;

(vi)   stores for samples, equipment, chemicals and glassware.

The basic functions of the laboratory should preferably be in separate rooms or defined parts of the main laboratory area, for example:

(i)   cleaning of glassware and equipment;

(ii)   sterilization of used glassware and incubated culture media;

(iii) preparation and sterilization of culture media;

(iv)   inoculation of culture media;

(v)   incubation of cultures;

(vi)   reading of results and subculture of micro-organisms;

(vii) interpretation of results and preparation of reports.

Thus, under ideal conditions a microbiology laboratory should be a suite of separate rooms rather than an all-purpose room. However, this cannot always be achieved and if only a single room is available, care must be taken to ensure that the flow of work is such that 'clean' and 'dirty' areas can be differentiated. Functions such as decontamination of pathogenic materials, storage of incoming samples and animal housing should not be attempted in a single room.

When planning the layout of a laboratory, attention must be given to routines or internal traffic patterns of the laboratory to minimize movements and to permit efficient performance of tasks. Provision must be made for the storage of protective coats adjacent to the laboratory area.

With regard to the fabric of the laboratory, the aim is to provide surfaces which are smooth, impervious and easy to clean and, additionally, for workbenches, which are resistant to heat and chemicals and free of cracks or exposed seams where micro-organisms could harbour and grow. Workbenches should be adequately equipped with gas, vacuum facilities, electricity, distilled water and hot and cold tap water. The workbenches should not be used to store laboratory equipment, media or other items. The space underneath may be used to accommodate cabinets and drawers. Side benches should be available for laboratory equipment, but equipment that produces vibrations, such as water baths, should be on benches separate from more delicate equipment such as balances and microscopes.

The ventilation system should have the capacity to draw off excess heat generated by equipment, but must not cause circulation of dust with the associated risk of contamination of samples and media. Microbiological laboratories should be regarded as a contaminated environment, thus the ventilation system must be designed to prevent air passage from laboratories to adjacent rooms. If equipment is installed to control the climate within the laboratory (air humidifiers, air conditioners, air dryers, cooling towers, ventilation trunking) it should be periodically cleaned to prevent build-up of micro-organisms which may directly infect laboratory personnel or contaminate equipment and media. Where incubators without built-in refrigeration are used, it is important that laboratory temperatures are stabilized (21–23°C) to allow such equipment to function efficiently. Maintenance of such a temperature is also essential to ensure the integrity of heat-sensitive media and reagents and prevent loss of viability of stock cultures.

Lighting should be good but not glaring. Reading of some tests requires a dimmed light, while for others natural daylight is best. Artificial lighting closest to day-lighting should be chosen. It may be necessary to place solar protection on windows; ideally this should be on the outside but inside is also acceptable. Media, chemicals and reagents should be stored in areas protected from direct exposure to sunlight, as performance may be altered. Similarly, analytical work should not be performed in direct sunlight as there may be an effect on final results.

3.4. Safety

In order to create a safe working environment in a microbiology laboratory, care must be taken to provide conditions which prevent infection by pathogenic organisms, minimize the risk of accidents and prevent contamination of samples. Procedures for safe working must be well defined and documented and a designated safety officer should be included in the laboratory organization. Each laboratory should have a safety committee which includes representatives of all grades of staff and which meets regularly with meetings minuted and actioned. Legal requirements in relation to health and safety at work must be implemented. All staff must be aware of the procedures for dealing with laboratory accidents, to whom they should be reported and how they should be dealt with. Protocols are needed for dealing with accidents of differing severity and hazard and there should be a ready supply of materials, equipment, protective clothing (aprons, gloves, face-masks, goggles), disinfection and neutralizing chemicals ('spill-kits') to contain spillages of both a microbiological and chemical nature.

New staff should receive induction training on laboratory safety on commencement of employment, while existing staff should receive regular updating of their safety training.

Staff should be aware of the need for containment facilities for particular groups of hazardous organisms, the procedures involved in the use of such facilities and the limitations of access. There should be defined written protocols for each level of containment appropriate to the biohazard classification of the particular group of pathogens. Most routine food and water microbiological examination can be performed under Risk Group 2 (moderate individual risk, low community risk) conditions unless the examination is aimed at isolating organisms in a higher biohazard classification (e.g. Risk Group 3, high individual risk, low community risk).

Safety audits, both internal and external, should be conducted at defined intervals, e.g. twice a year internally, annually externally. The importance of following the laboratory safety code should be stressed to all staff. The first aim is personal protection and safety, but protection of the product in the interests of producing reliable test results must be included. Each laboratory should also have written plans to cover fires, fire-fighting and evacuation of premises if an emergency occurs. Annual training on response to fires should be provided for all staff. Safety equipment, including fire extinguishers and fire hoses, should be checked regularly to ascertain that they are in their proper places and functioning correctly.

Assessments should be made of the use of hazardous chemicals. This is often a legal requirement. EU Member States should have in place regulations consistent with the provisions of the directive on the protection of workers from the risk related to exposure to chemical, physical and biological agents at work (Council of the European Communities, 1980: Directive No 80/1107/EEC). Under such regulations, written assessments should be produced for all procedures which require the use of chemicals, reagents, media or microorganisms and should be regularly updated, new ones being prepared if a procedure is modified or a new one introduced.

Examples of the important points in laboratory practice relating to safety are:

(i)   access to laboratory limited to authorized persons only;

(ii)   mandatory wearing of laboratory coats within the laboratory area;

(iii)   wearing of laboratory coats and other protective items not permitted outside the laboratory area, particularly in public areas or where food is being consumed;

(iv)   use of additional personal protection (gloves, goggles, masks, etc.) when required for a specific activity;

(v)   windows and doors closed while working;

(vi)   no pipetting by mouth, correct use of pipetting aids;

(vii)   no eating, drinking or smoking within the laboratory area;

(viii) regular disinfection, at least daily, of workbenches;

(ix)   prevention of aerosol production when handling microorganisms;

(x)   effective disinfection/sterilization of laboratory waste or glassware before cleaning or re-use, or removal (unless incineration is under laboratory control when the sterilization step may be omitted);

(xi) transport of contaminated items (cultures, media, waste) in sealed leak-proof containers;

(xii) prevention of strong air currents in the laboratory area;

(xiii) appropriate use of fume cupboards, inoculation cabinets, laminar flow cabinets, to protect the product and reduce the risk of human infection;

(xiv) provision of separate containers for the disposal of broken glass, laboratory sharps (syringes, etc.) and solvents and other hazardous wastes.

3.5. Working procedures

Each laboratory should have manuals of methods which cover all the procedures undertaken in the examination of food and water. These should be followed strictly with no deviation unless authorized by the appropriate senior member of staff. Similarly, there should be written procedures for the use of both everyday and specialized equipment, the care of, cleaning and maintenance of such equipment and the actions to be taken when there are deviations from normal operation. All written protocols should be readily accessible to the staff concerned. Methods should, wherever possible, be up-to-date and established as standard. Where other methods are used they should be suitable for the purpose concerned, clearly documented and, if possible, agreed with the client.

3.6. Communication and customer relations

Within the laboratory, there should be good communication between staff at all levels and between levels. There should be clear channels of communication from director/laboratory manager right through to support personnel.

Laboratories should maintain a good working relationship both within a department and with other units/departments with which there is a working arrangement, and also with suppliers of equipment, reagents, etc. and particularly with customers.

Procedures for dissemination of results should be defined, with clear allocation of responsibility for the giving out of results. Means of rapid communication of results to customers when findings are unusual, such as the isolation of a pathogen or an unexpectedly high level of a particular organism or group of organisms, should be utilized (e.g. telephone, fax, electronic mail). It is also important to identify at what stage results which may affect public health are notified to enforcement authorities, governmental departments, etc.

It is essential that good records are kept of laboratory results. All observations, calculations and other information of practical relevance to the tests performed should be recorded and signed or initialled by the member of staff involved. The information recorded should be sufficient to enable the detection of any sources of error and for the test to be repeated if required. There should be complete traceability in the whole examination procedure from receipt of sample to transmission of final results. Records of laboratory examinations should be archived in a well-organized manner, protected from loss, damage or misuse and retained for a minimum period of six years.

3.7. Complaints management

Laboratories should have a defined procedure for the investigation of complaints received from clients or other parties and any anomalies identified in relation to the activities of the laboratory. Records should be kept of any such complaints or anomalies and of any action taken. This is of particular importance in the event of possible future legal proceedings. A member of staff receiving a complaint or identifying an anomaly should refer the matter to a senior member of the technical staff who will deal with the matter, involving other senior members of staff as appropriate, and take corrective action. All complaints received relating to the laboratory's activities should receive a written reply.

Anomalies which might have an effect on laboratory results include fluctuations in incubation or refrigeration temperatures outside the established accepted ranges, media not conforming to specifications such as pH, etc. Inclusion of controls with daily tests will allow the laboratory to determine whether such anomalies have an effect on the final result, in which case the test will be declared void and the customer informed, or whether the results can stand.

It is also important to measure customer satisfaction with the service provided by the laboratory. This can be done by the use of written questionnaires (which tend to be rather impersonal) or by the establishment of liaison groups where laboratory staff and customers can meet face to face and discuss any problems or plans for future work.

References

Anon., 1984. Report of Task Force D at the International Laboratory Accreditation Conference, London, UK, Department of Trade and Industry, London, UK, cited by FAO (1991) below.

Council of the European Communities, 1980. Directive No 80/1107/EEC on the protection of workers from the risk related to exposure to chemical, physical and biological agents at work. Official Journal of the European Communities, 1980, No L 327, 3.12.1980, p. 8.

Other sources of information

Garfield, F.M. ,1991. Quality Assurance Principles for Analytical Laboratories. Association of Official Analytical Chemists, Arlington, VA.

Food and Agriculture Organization, 1991. Manual of Food Quality Control. 12. Quality Assurance in the Food Control Microbiological Laboratory. FAO Food and nutrition paper 14/12. FAO, Rome.

Nordic Committee on Food Analysis, 1989. Handbook for Microbiological Laboratories. Introduction to Internal Quality Control of Analytical Work. Report No 5, Technical Research Centre of Finland, Espoo, Finland.

Netherlands Draft Standard, 1993. Bacteriological examination of water. General principles for quality assurance of bacteriological examination of water. Draft standard NPR 6268, 1993.

World Health Organization, 1993. Laboratory Biosafety Manual, 2nd edn. WHO, Geneva.

Chapter 4

Sampling

4.1. Introduction

The quality of the final report issued by a laboratory will be influenced by the quality of the sample taken and submitted for analysis. Often the analysis request will require an answer to specific questions and therefore the correct sample must be analysed by the most appropriate techniques. Therefore, samples should be taken within an agreed sampling plan following discussion with the laboratory user about the questions to be answered. Then the type, quantity and frequency of sampling can be agreed.

Within a complete quality assurance programme, the careful control of the sampling process is important. Measures taken in controlling the analytical process may become void if major errors are introduced at the sampling stage. These can involve faulty sampling procedures, transport and documentation.

It is important that every single factor that may influence the sampling process and ultimately the final result is identified and understood. In this way, the relative importance of these factors can be assessed and control measures can be introduced. These factors are illustrated in the sampling process (Fig. 4.1).

Three identities are involved in the sampling process, the sampler, the receptionist and the analyst, but they may be represented by a single person. The sampler influences the majority of the factors and particular attention will have to be paid to his/her training and

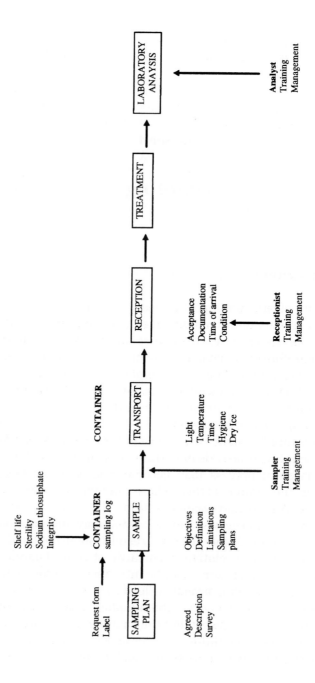

Fig. 4.1. The sampling process — sources of imprecision.

management. Protocols will need to be in place to control the important factors and these can be produced from the information given in this chapter.

Although the sample should spend very little time in transport and in laboratory reception, and should be handed over fairly quickly to the analyst, care must be taken to ensure valid documentation and correct entry onto the laboratory record to avoid delays. Samples will then give true information about the food or water from which they were obtained and an audit trail will allow the correct interpretations to be made and necessary actions taken. This documented history of the sample becomes very important if results are ever disputed.

4.2. Distribution of organisms in food and water

A sample is a small proportion of food to be examined or a very small volume of the water being investigated. The dispersion, species and number of micro-organisms in the sampled fraction should reflect those in the whole batch of food, or in the water system from which it has been taken. This assumption is true only when the consistency and the processing technique allow for a homogeneous distribution of the micro-organisms within the food and among the food units, e.g. bottled pasteurized milk may approach this distribution.

The distribution of micro-organisms in food and water is not homogenous and has been described as random or contagious. In random distribution, any one individual cell is randomly located in the mass of food or water. In contagious distribution, the cells may occur in clumps or aggregates with an abnormal spatial dispersion. Experience shows that the most frequent distribution type is the contagious (Shapton and Shapton, 1991).

Homogeneity is not a quality as such but can be quantified; it depends on the number of particles per unit of material and on the sample size. Usually, it is expressed as part of the uncertainty of measurement and in a homogeneous sample the participation of this uncertainty towards the uncertainty of the measurement method is negligible. If not, the sample size may be adjusted or the number of replicate measurements increased.

A series of factors may influence the spatial distribution of organisms, such as the flora of raw materials, mincing, blending and handling techniques, the storage temperature, cooking and chilling. Contamination during manipulation, followed by storage at ambient temperatures, may cause the growth of organisms resulting in development of micro-colonies.

Foods with multiphasic systems (e.g. meat and gravy), in which each phase may have a different flora and microbial distribution, require multiple sampling for a composite analytical answer. In this case, the ingredients of each phase should be tested separately. It is obvious that the type (pathogenic or not) and the dispersion of organisms in food will be reflected in the analytical data and will affect the sampling programme used for the analysis (Jarvis, 1989).

The objective of the sampling programme is to control any uncertainty that may arise as a consequence of these organism distribution patterns.

4.3. Sampling plans

Sampling is not just the procedure of taking a small proportion of food or water for analysis. It aims to provide information on the microbial characteristics of the food or water, which on the basis of agreed criteria may help to accept or reject a batch of the food undergoing processing or in storage, and to provide a health hazard or hygiene assessment. The objectives of the sampling will determine the details of this procedure, e.g. the size of the sample, the frequency of sampling, the point of collection and the time of sampling.

The need for a specific programme on what, when and how often samples should be taken, the parameters to be determined and the criteria for final decision have resulted in the development of sampling plans. Sampling plans for water may be described in EU Directives such as the bathing water directive (1975), the surface water directive (1979) or, in the case of drinking water, may be based on the size of the population supplied or may be related to the volume of water produced at a treatment works.

The International Commission on microbiological specifications for foods (ICMSF, 1986) states that "a sampling plan is a procedure for withdrawing a sample, carrying out analysis of appropriate units and making appropriate decisions based on agreed criteria". A similar definition, but one adapted more to the food industry is quoted by Messer *et al.* (1982). Some guiding points for developing sampling plans are health hazard assessment, keeping quality of end products, trend analysis, checking of raw materials, control of production line, environmental hygiene or a food poisoning outbreak.

The assessment of food safety and potential health risks usually follows the principles of hazard analysis critical control point (ICMSF 1986, *Codex alimentarius*, 1982). The rationale of sampling plans is then based on these identified critical control points.

For more details, discussion and description of sampling plans, the publications of ICMSF (1986) and Compendium of methods by APHA (Vanderzant and Splittstoesser, 1982) may be used.

4.4. Food

Sampling of food can be done for many reasons. The following paragraphs give details of the various categories of sampling.

4.4.1. *Acceptance /rejection testing*

Acceptance /rejection testing of foods in international trade is done on the basis of agreed specifications and sampling plans. A full explanation of the principles and specific applications of sampling foods for microbiological analysis is given in ICMSF (1986). Testing for pathogens is often carried out on a presence/absence basis in defined amounts of sample. Where results are expressed in terms of plate counts it is usual to allow some latitude in results that marginally exceed the desired maximum count.

4.4.2. *End product testing*

End product testing can be applied by the food producer to document that the food meets legal or internal quality standards. For this, the relevant national or EU limits are applied to produce sampling plans, or individual specific internal limits, and sampling plans can be designed to meet the perceived requirements. Sample numbers typically may vary from one to five per batch. Some countries do not specify the number of samples to be taken (e.g. the Netherlands), whereas this can be fixed in other countries (e.g. France) who apply an ICMSF-type scheme where $n = 5$. The EU sampling schemes for milk and milk products are also ICMSF-type schemes with $n = 5$ (Council Directive 92/46/EEC, 1992).

4.4.3. Trend analysis

Trend analysis is a different way of sampling and testing; accumulated data are used to measure changes in the product rather than in a specified batch. Most food products are made following a process in which microbiological hazards are considered to be under control, based on a previous HACCP study (hazard analysis critical control point). Such a HACCP study should follow the seven principles detailed by *Codex alimentarius*, i.e. hazard assessment, critical control point identification, establishing critical limits, monitoring procedures, corrective actions, verification and documentation procedures (FAO/WHO *Codex alimentarius* Commission, 1993). In this approach, known microbiological hazards are controlled at specific points in the process by agreed procedures and product/process specifications. Microbiological sampling and analysis can then be restricted mainly to monitoring and verification. The resulting product will be determined to be of good quality when the process is running within agreed limits. In these situations, a number of non-microbiological process parameters can be monitored to keep the process in control. To check that the product is meeting its specifications or to see whether there is any drift in microbiological quality, a limited number of samples is collected during the year.

4.4.4. Statutory testing

Statutory testing is performed when a control authority wants to check whether a batch of food or food ingredients meets the legal requirements for such food. The sampling by the food inspector should be carried out according to the requirements laid down in regulations. When such sampling takes place at the food manufacturing plant or in a shop, the representative of the manufacturer/shop is informed and duplicate samples are collected and made available for independent analysis by the manufacturer or shop owner. In particular situations, an additional dummy sample can be collected and used for temperature measurements during transport.

4.4.5. *Investigative sampling*

Investigative sampling can be carried out by a food manufacturer when the source of contamination of a product by pathogens or spoilage organisms needs to be determined. The sample size numbers and sample position will be dictated by the nature of the problem. In such investigations, the number of samples may rise by several hundreds. When raw material contamination is suspected to be the origin of the problem, the sampling will focus on the types of ingredients used. Sampling of intermediate products in a food plant may also be required. Investigations might also be focused on samples which have been incubated initially for a longer period in order to increase the contamination.

4.4.6. *Outbreak analysis*

Outbreak analysis requires a type of sampling which focuses on known or suspect foods which could have caused a problem. In outbreaks of food-borne illness caused by *Salmonellae*, where it is necessary to obtain conclusive evidence that a particular incident was caused by a certain product, thousands of samples may have to be collected and analysed. Where the source of an incident is unknown, use is often made of a case control study in which the products eaten by the cases are compared with products eaten by matching non-ill controls. These studies can indicate which food product is more likely to have caused the illness and sampling can then be focused on these foods.

4.4.7. *Environmental and hygiene monitoring*

For hygiene evaluation and environmental sampling, the sample size and numbers mostly depend on the objective of the investigation. When the source of a spoilage organism has to be traced in a food production plant, the samples can be taken from food ingredients, intermediates and products, but it is helpful to take swab samples or scrapings from plant and equipment. When the food processing environment is monitored for contaminants or pathogens, this can be achieved by using small sponges (maxi swabs) or by collecting food debris or dust.

When the hygienic status of the food plant is monitored, agar contact plates are often used (Rodac plates, dip slides, agar sausages) to sample the flat surface of equipment. Recently, the use of tests to determine residual adenosine triphosphate (ATP) has been proposed as a quick verification of the cleanliness of equipment, indicated by absence of microbial or food ATP.

Since agreed sampling protocols for hygiene/environmental monitoring do not exist, protocols will have to be introduced for specific investigations.

4.4.8. Food sample transport

Once the sample is collected it should be transported hygienically under strict conditions of time and temperature, to preserve its microbiological status. It should be accompanied by sample identification forms. A well-documented procedure should be used to control the transport and prevent damage to the samples. When a frozen sample arrives defrosted, or a chilled sample is received at ambient condition, the interpretation of the results is difficult, and it is possible that it should not be analysed. When the aberration is less apparent but the microbiological contamination may be significantly affected, there may be a false analytical conclusion. An example of this is the temporarily freezing (by solid carbon dioxide) of wrongly chilled samples which may eliminate the freeze-sensitive vegetative cells of *Clostridium perfringens* or reduce the contamination of other enteric pathogens. Such samples could arrive at the laboratory at the correct chilled temperature and appear unharmed by the transport. Packaging and transport conditions should therefore be carefully specified to assure that the microbiological condition of the sample is not significantly changed.

In some situations more relaxed conditions may be acceptable. This might be the case for presence/absence testing of spoilage bacteria in environmental samples or with sampling of rather stable foods in which only slow growth will take place. The transport conditions may allow for a wider time/temperature span.

Time and temperature

The conditions for transport for most ambient stable or frozen foods can be identical to the storage conditions. This means transport of frozen products at –20°C and for ambient stable products transport at 20–30°. The transport of perishable fresh or chilled food samples is more difficult.

Transport of fresh foods like vegetables at ambient conditions will change the microbial load unless the time for transport (i.e. from sampling to analysis) is very short, typically two to four hours.

For all perishable products, chilled transport is preferable, although even chilling can affect very sensitive microbial cells through cold shock.

For already chilled food samples, chilled transport evidently is the recommended route. The maximum temperature during transport should be specified. The analysis should be carried out as soon as possible on arrival, but where the determination cannot be performed within six hours, chilling down to 2–10°C could extend the time between sampling and analysis to 24 hours. FAO recommends an analysis within a maximum of 36 hours when samples have been chilled to 0–4°C (Andrews, 1992). For all chilled products, maximum transport time/temperature combinations should be specified in the sampling protocol. Time/temperature indicators or electronic data loggers can be shipped with the sample to monitor or record the actual temperatures during transport.

Hygiene

The hygienic status of the sample container, of the refrigeration medium in the case of melting ice and of the secondary packaging should be verified to limit contamination from outside the sample.

The sampler should record the time of sampling and the temperature of the sample, particularly for fresh samples to be chilled and already chilled samples. In combination with time of arrival and the measurement of the arrival temperature, a record is created which can document the integrity of the sample transport step.

4.4.9. *Intermediate storage of food samples*

In general, the same principles apply for intermediate storage as for sample transport. As soon as the sample arrives at the laboratory its general condition should be noted, including the temperature of the container, when transported refrigerated.

If possible, the sample should be examined immediately after receipt. If analysis must be postponed, however, the sample should be stored in such a manner that it is protected from any chemical, physical and mechanical factors which may cause changes in the sample, leading to significant microbiological multiplication or death. Recommendations are:

(i)   Store a frozen sample at –20°C until examination;

(ii)   Store a non-perishable, canned, low-moisture or otherwise shelf-stable sample at room temperature until examination. Avoid temperatures above 40°C. A canned product should be refrigerated when there is a risk of 'blowing';

(iii)   Store an unfrozen perishable sample refrigerated between 0 and 4°C. Do not store for longer than 36 hours, including transport time;

(iv)   In principle, a perishable sample should not be frozen, but if examination cannot be carried out within 36 hours, freezing of the sample may be considered. It should be realized that such a treatment might affect the microbiological status of the sample; for this reason it should always be reported. The freezing of a sample should be done rapidly (within 2 hours);

(v)   The laboratory should be equipped with sufficient freeze, cold and other stores to keep the samples under the proper conditions.

4.5. Water

There are several different categories of water sampled to achieve different objectives; these should be clearly stated in the sampling strategy. This should be well thought out, understood and agreed by all personnel involved in the process. In some situations, sampling is not under the control of the examining laboratory and it will be necessary to produce clear instructions of the techniques to be used. In these situations, interpretations of results may be difficult if the sample history is not known. Training of samplers will play an important part in achieving good quality samples.

If a systematic sampling programme is used, e.g. to describe a situation or detect a trend, it is prudent to check that the frequency of sampling does not coincide with one of the natural cycles of the system (e.g. tides). Conversely, if the study aims at detecting such cycles and their influence, it is necessary to begin with frequent sampling.

4.5.1. *Recreational water testing*

In any monitoring programme, the objective is to obtain samples which are representative of the whole body of water. This presents a difficulty because of the lack of homogeneity and the effect of unpredictable inputs of storm water run-off and variable effluent inputs. It is further complicated by the effects of tide, sunlight and temperature and local water circulation patterns. To obtain a representative sample, attention must be paid to the selection of the sampling station, strict adherence to proper sampling procedures, the identification and labelling of the sample and the prompt transport to the examining laboratory. Protocols for all these aspects will need to be agreed and written. It should be remembered that sampling at different sites at the same time from the same body of water will produce quite wide variation in the parameters measured. The ultimate objective of a monitoring programme is to observe trends; a single sample will not give a complete picture of recreational water.

Sampling sites

The sampling sites should be chosen to reflect exposure of bathers taking into account that most children are exposed in areas close to the water line. In these areas, there may be considerable concentrations of sand and sediment particles and sampling sites are thus commonly established where the water depth ranges from 1.0 to 1.5 m. A base-line survey of water quality in a recreational water mass provides a sound basis for establishing the location and number of sampling sites. For recreational water monitoring programme, detailed local investigations on currents, tides, volumes, types and sites of discharging effluents and of prevailing winds will help to determine sampling sites. A base-line sampling survey of water quality may lead to better selection of sampling sites to reveal particular patterns of water quality deterioration and final selection of sampling stations for bathing or shellfish cultivation areas. EEC Directive 76/190 (1976) recommends that water samples should be taken at 30 cm below the surface; this avoids any possible interference due to particulate or floating material. The sampling frequency should be directly related to the intensity and temporal pattern of recreation activities. EEC Directive 76/160 (1976) specifies a minimum requirement of one sample every 15 days, but higher sampling frequencies are useful, especially where tidal currents induce rapid variations in quality. In such situations, a high sampling frequency is better than a large number of sampling points.

Recreational water sample collection

It is important to ensure that samples do not become contaminated at the time of collection or before they are analysed. The basic procedure for collecting a sample is to hold the bottle near its base, to place it below the surface of the water and to remove the cap so that it can be filled with water at the desired depth. As the bottle fills, it should be pushed gently forwards through the water to prevent contamination from the sampler's hands. When sampling using an extension arm, the bottle can be introduced open and upside down below the surface of the water and then turned so that the bottle fills at the desired depth. When sampling from a boat, the sample should be collected from the upstream side of the boat to prevent contamination. The bottle should not be filled completely; a small airspace should be left so that adequate mixing by shaking can be carried out.

4.5.2.   *Potable water testing*

Water samples are tested at several points in the water production process. The microbiological quality of raw water and any changes in the microbial load provide important information for water treatment operational decisions. Water is also sampled when it leaves the treatment works and also when it is in the distribution system and will include samples from customers' taps.

Raw water sample collection

These samples are often taken from taps on the raw water supplies to a treatment works. The taps must be cleaned before sampling (see below). Some taps are kept open so that there is a continuous flow. Care must be taken to prevent contamination of the taps and the sample. Some raw waters are chlorinated before entering the treatment works and residual disinfectant must be neutralized in the sample bottle. In certain situations, chlorination is carried out at the base of a bore-hole and it is therefore difficult to obtain a representative sample without chlorination. If the chlorination is stopped, it may expose the users of the supply to a health risk and therefore it should not be stopped without taking precautions to protect the consumers.

A protocol will need to be written for each sampling site to obtain consistency of quality.

Sampling from the distribution system

The frequency of sampling is laid down in national or European regulations and is based on the cost effectiveness of testing at a level to ensure public health. The samples are taken from a mixture of fixed sampling points; reservoirs and hydrants and a random selection of customers' taps. Hydrants must be disinfected and must be allowed to run before collecting the sample. This is not an easy task and should follow a strict protocol. It will take about 20 minutes to obtain a good-quality sample.

Disinfection of taps

Taps can easily become contaminated from the environment and the inner parts may be affected by growth of micro-organisms in a biofilm. It is therefore important that the extension of the tap is adequately cleaned and that the tap is run for long enough to flush out any organisms in the tap mechanism.

Metal or ceramic taps can be disinfected by flaming with a blow torch. Taps made of other materials will not withstand heating and must therefore be disinfected using 70% alcohol which is allowed to evaporate, or by using a hypochlorite solution of 1 g/l. Following disinfection, the tap must be run for three minutes to flush out organisms in the tap mechanism and to ensure that no residual disinfectant enters the sample bottle.

Samplers must be trained and must understand the factors that can influence the quality of the sample.

4.5.3. Bottled water testing

The regulations relating to mineral waters and water in containers (bottled waters) require control of microbiological quality of the water source and the end product within 12 hours of bottling. Samples from bottles, caps, the environment and machinery will often be required. If these waters are sampled in the distribution chain after the 12-hour limit, the viable count may vary considerably. The viable counts in freshly bottled waters are usually very low. When sampling from bottles, the integrity of the cap is essential and the water should be poured into the analysis vessel direct. It may be necessary to de-gas and neutralize carbonated waters. De-gassing can be achieved by using a vacuum pump and filter to withdraw the gas from the water while it is being stirred with a magnetic stirrer. Neutralization can be carried out using sodium hydroxide.

4.5.4. Other water testing

Samples from swimming pools, hydrotherapy pools, spa pools and whirlpool baths are taken at varying frequencies to measure the microbial contamination and the adequacy of disinfection procedures. Baseline studies are of considerable value for the determination of sampling sites and frequencies, and the effects of other factors such as user density on the microbiological quality.

A major factor that can influence the quality of the results is the effect of residual disinfectant. This must be neutralized with the appropriate neutralizing agent for the disinfection system in use (see Chapter 6, Section 2.5).

4.6. Sample records

It is essential that samples are clearly labelled just before their collection and that the labels are clear and cannot become detached from the sample. The samples should be taken in accordance with a pre-arranged plan and a sampling log should be completed by the sampler, giving details of date, time, identity of the sampler, weather conditions and any abnormal circumstances. If each sample bears a unique number, the history of any sample can be traced through these records.

4.7. Sampler training

It is important that the sampler understands the objectives of the sampling programme, the methods of sampling and all factors that can influence the quality of the sample. The sampler carries a considerable responsibility and will need a properly structured training programme. The sampling process should be audited at intervals by the laboratory technical manager and quality manager.

4.8. Sample transport

During transport, the death of the sample, or conversely the overgrowth of micro-organisms in the sample must be prevented. To achieve this, the sample must be protected from ultraviolet and visible light and high temperatures. This is usually achieved by using cool-boxes containing ice packs. In temperate climates, the ambient temperature may not affect the organism count. The transport system used, however, should be validated and documented; this can be achieved by the use of recording thermocouples.

The sample should be analysed as soon as possible. In some countries, time-limits may be laid down in regulations.

It is important that the samples be transported in hygienic conditions so that cross-contamination is avoided and that dirty samples should not be mixed with clean samples. The sampler will need to know when to wash or rinse his hands to avoid gross contamination of the outsides of clean sample bottles.

4.9. Sample reception

This is the last important part of the sampling process before analysis. A protocol for reception will be necessary, which should state the persons who are permitted to receive samples and the checks they should make. These are typical first-line checks.

Control of the sample accompanying the form:

(i)   Has the correct sample been taken?

(ii)   Has the form been completed correctly?

(iii)   Is the name of the sampler known?

Condition of the sample:

(i)   Has the sample been cooled?

(ii)   Is the label intact and readable?

Control of the sample bottles:

(i)   Has the correct sample bottle been used?

(ii)   Is it properly closed?

(iii)   Is the closure leaking?

(iv)   Is the volume sufficient?

The receptionist should record the time of receiving the sample on the form. The time when the analysis commences should be recorded later by the analyst.

4.10. Summary

It will be seen that at this sampling stage many factors can have a great influence on the results obtained. If attention is given to sampling protocols and sampler training, with a thorough understanding of the whole process, then high-quality samples will enter the laboratory and the results will be meaningful. In situations where the laboratory does not influence the sampling process, organizational changes should be requested by the analyst and sampling protocols should be made available to all who wish to use the laboratory's services.

References

Andrews, W., 1992. FAO Manual of Food Quality Control, 4 (Rev. 1), Microbiological Analysis. FAO, Rome, Chapter 1, Food sampling, pp. 1–8.

Council Directive 76/160/EEC of 8 December 1975 concerning the quality of bathing water. Official Journal of the European Communities L 31, 5 February 1976, p. 1.

Council Directive 79/869/EEC of 9 October 1979 concerning the methods of measurement and frequencies of sampling and analysis of surface water intended for the abstraction of drinking water in the Member States. Official Journal of the European Communities L271, 29 October 1979, p. 44

Council Directive 92/46/EEC of 16 June 1992 laying down the health rules for the production and placing on the market of raw milk, heat-treated milk and milk-based products. Official Journal of the European Communities L268/1, 14 September 1992, pp. 1–32.

FAO/WHO *Codex alimentarius* Commission, 1993 Report of the 26th session of the Codex Committee on food hygiene, Alinorm 93/13A, Appendix II, Guidelines for the application of the hazard analysis critical control point (HACCP) system, pp. 26–32.

ICMSF, 1986. Micro-organisms in Foods, Vol. 2. Sampling for Microbiological Analysis: Principles and Specific Applications, 2nd edn. Blackwell Scientific Publications, Oxford.

Jarvis, B., 1989. Statistical Aspects of the Microbiological Analysis of Foods. Elsevier, Amsterdam.

Messer, J.W., Midura. T.F. and Peeler, J.T., 1982. Sampling plans, sample collection, shipment and preparation for analysis. In: C. Vanderzant and D. F. Splittstoesser (Eds.), Compendium of Methods for the Microbiological Examination of Foods, 3rd edn. APHA-USA.

Shapton, D.A. and Shapton, N.F., 1991, Principles and Practices for the Safe Processing of Foods. Butterwarf and Haynemann Ltd.

Vanderzant, C. and Splittstoesser, D. F. (Eds.), 1982. Compendium of Methods for the Microbiological Examination of Foods, 3rd edn. APHA-USA.

Chapter 5

Equipment

5.1. Introduction

The requirements for the production of high-quality laboratory results are:

(i)   well-trained and experienced personnel;

(ii)   well-maintained and functioning equipment suitable for each type of test;

(iii) well-documented procedures;

(iv) a good-quality sample.

Even the best personnel will not be able to produce quality results if the equipment available is not suitable or is not functioning correctly. For example, sterilization of culture media in an autoclave which does not maintain the correct temperature will result either in non-sterility or in the alteration of heat-sensitive components of the culture media. Similarly, efficient isolation or enumeration of *Salmonellae* and *Escherichia coli* can be achieved only if the incubator or water bath maintains the correct temperature (i.e. 42°C and 44°C) within acceptable fluctuations.

It is obvious that the accuracy and precision of bacteriological results depend not only on the competence of laboratory personnel and methods used, but also on the use and performance of the specific apparatus.

Table 5.1

Equipment used in food and water microbiology laboratories

Sterilizing equipment	Membrane filtration apparatus
Autoclave	Balances
Media preparation	Deionizers
Pressure cooker	Distillation apparatus
Hot air oven	Refrigerators
Filtration	Freezers
Incubators	Anaerobic incubation
Water baths	Homogenizers
Thermometers	Colony counters
pH meters	Microscopes
	Automated ELISA equipment

Thus any quality assurance programme should include monitoring the performance of equipment and promoting adequate maintenance, with an aim of:

(i)   using/purchasing/validating appropriate apparatus for each test procedure;

(ii)   maintaining and checking regularly its performance when in use.

A list of the most important apparatus required in food or water microbiology laboratories and some of the technical specifications is given in Table 5.1. Each item of equipment is dealt with in separate sections of this chapter. Following a description of principles and function, the methods of validation of good performance are given under the headings calibration, validation, use, checks, recording, maintenance and safety where these are appropriate.

5.2. Types of sterilization apparatus

The purpose of sterilization is to inactivate or remove all living organisms (not only vegetative cells and viruses but also spores) and can be achieved by known methods such as moist heat, dry heat, filtration, ionizing radiation and by some chemicals such as ethylene oxide. Sterilization is always dependent on the time of exposure to the selected method and the quality assurance programme should give in detail the checks necessary to ensure that sterilizing conditions are achieved for the correct length of time on each occasion of the operation.

In food and water microbiology laboratories, sterilization apparatus is usually autoclaves for moist heat and hot-air ovens for dry heat. The autoclaves now available have developed considerably and have automatic process controls. Ionizing radiation and ethylene oxide sterilization apparatus, because of their inherent greater toxicity, are more complex and are usually situated at manufacturer's premises, where they are used in the production of sterile heat-sensitive products.

5.2.1. *Heat sterilization*

Heat is applied moist (steam) in autoclaves or dry in hot-air sterilizers. Moist heat is more efficient than dry heat and therefore takes less time. For both types of sterilizers the following principles should be considered in relation to successful sterilization:

(i)   The micro-organisms should be in direct contact with heat and, if protected by layers of organic material or within firm packaging, will need time for the heat to reach them.

(ii)   The organisms are not inactivated instantly. They must be exposed for a period of time at a specified temperature. The time allowed for sterilization corresponds to the time the materials are at sterilizing temperature and does not include penetration time or the time necessary to reach the sterilizing temperature.

It is necessary to validate the process using standard loads and thermocouple recording of temperatures attained in the various parts of the load.

5.3. Steam sterilization

For steam sterilization a steam sterilizer or autoclave is required. Applications include:

(i)   sterilization of media and other bacteriological equipment, e.g. laboratory glassware, diaphragm filters and holders, blending apparatus, etc.;

(ii)   sterilization (decontamination) of contaminated materials.

Steam sterilizers are chambers fed with pure steam via a control valve to achieve and maintain a specific pressure/temperature for a required time period. There are many types of steam sterilization equipment and the main problems associated with them are air removal, superheating, load wetness and damage to material.

Air removal is the first important step; it depends on the type of sterilizer and can be influenced by uneven loading. The method of air removal varies with the type of steam sterilizer but may be achieved by one of the following:

(i)   dilution by mass flow;

(ii)   gravity displacement;

(iii)   pressure pulsing;

(iv)   high vacuum;

(v)   pressure pulsing with gravity displacement.

In conventional vertical autoclaves, a quantity of water is heated up in the boiler and the air is driven out by a prolonged flow of steam. The point at which only pure steam is achieved may not be very clear. Also, a lengthy period is required in order to heat the water and afterwards to cool it down, which can result in prolonged exposure to heat and often in excess wetting of the materials to be sterilized.

In an autoclave with a separate steam generator, the air is driven out more quickly and effectively. It has the advantage that the media can be cooled more rapidly after sterilization to less than 100°C by reducing the pressure with compressed air compensation. Any condensate resulting may, if necessary, be removed by vacuum. An alternative is the double-walled autoclave, usually with a vacuum phase to remove air before the introduction of steam and sometimes with pressure pulsing to obtain more efficient air removal.

The loading of the autoclave may affect the time/temperature relationship for those items, e.g. flasks, containing media and additional time may be necessary. Wrapped items should be loosely packed to allow steam/heat penetration. The sterilizing process should therefore be validated using standard loading patterns.

5.3.1. *Temperature and pressure measurement and recording*

Autoclaves vary in complexity from the simple pressure cooker to the sophisticated microprocessor- controlled autoclave capable of achieving different types of cycles applicable to different types of loads. Regardless of the type of autoclave used, its performance will depend on temperature/pressure measurement and control for a set period of time.

(a)   Two temperature measurement systems operating independently of each other should be provided, one of which is used for temperature control in the autoclave (if that regulating principle is applied). The second is used for recording the temperature by means of a floating thermocouple placed in the load. The ideal solution is for the autoclave to be controlled via the load temperature.

It must be possible to read the prevailing temperature by one or other of the two measurement systems during the process. The temperature scale must cover a range of 50–150°C and the total imprecision of the measuring system must be less than 1°C. This must be achievable in all circumstances.

The error on the temperature regulating apparatus must be less than 0.5°C. It must be possible to position the two temperature recorders in such a way that their readings can be compared. Recording on paper should be such that a scale value to 1 mm/°C is possible. A scale gradation in 2°C is preferable for a better readability.

(b)   Two pressure measurement systems operating independently of each other should be provided, one of which uses a measuring system for pressure regulation (if this system is used) and the other a recording instrument. Both systems must be resistant to high temperatures and belong to at least DIN Class 1.

The pressure scale must have a range of 0 to at least 350 kPa. Recording on paper must be such that a scale value of 0.25 mm/kPa is possible.

(c)   Paper speed must be at least 100 mm/h.

(d)   Pressure and temperature must be recorded on the same sheet.

5.3.1.   *Calibration*

A departure from the measured value by 1.5°C or less is permissible. Adjustment of the apparatus must be made by suppliers having calibration accreditation.

5.3.2.   *Validation*

Validation should be regarded as the basis for assessing the effectiveness and reproducibility of the sterilization process. Initial validation (start-up validation) is undertaken when the product sterilizer/ process combination is tested for the first time. Revalidation is necessary after work is done on the sterilizer or peripheral equipment that influences the process. Revalidation may also be necessary when deviations are noted within the course of the process that fall outside the tolerances.

Validation may be undertaken by means of:

(i)   physical measuring methods, when the progress of the physical parameters is checked (technical validation);

(ii)   microbiological measuring methods, when products with a defined contamination are used (microbiological validation).

The following points are of importance during validation:

(i)   identification of the sterilizer;

(ii)   specification of the measuring and control apparatus of the sterilizer;

(iii)   determining the temperature distribution in a loaded autoclave;

(iv)   checking the pressure measurement and pressure recording;

(v)   reporting.

5.3.3.   *Use*

If sterilization is to be reliable, the following information must be noted and laid down in procedures:

(i)   The steam used for sterilization must be saturated.

(ii)   All air must be removed from the sterilization vessel and from the load. Account should be taken of the fact that air is heavier than steam and that air and steam do not mix easily. This will affect the loading of the autoclave and the place where the temperature is measured.

(iii)   Packaging of the material must be sterilized.

(iv)   The autoclave should preferably be loaded with flasks having the same volume.

(v)   The volumes of the flasks may differ by a factor of no more than two. If the autoclave is used for decontamination, a larger difference is permissible.

(vi)   The progress of the temperature change must be monitored in at least one flask.

(vii)   If the flasks have different volumes, the progress of the temperature change must be monitored in the centre of the flask with the largest volume.

(viii) If steam is applied to the autoclave, the pressure must first be reduced and the steam must be able to cool.

(ix)   The steam supply pipe must be fitted with water separators.

(x)   The correct sterilization temperature must be maintained. This temperature must be checked by continuous measurement in the boiler and load.

(xi)   The autoclave should never be blown out with a forced draught. This will result in foaming, boiling over and slowing down of the heating of the media. Opening must be done carefully because of the steam escaping.

(xii) The load should be allowed to cool, and heat-resistant gloves and a visor used to remove the load from the autoclave.

5.3.4. Checks

The primary check on the operation of the autoclave is by examination of the temperature/pressure chart record of the particular process cycle to check that the correct physical conditions for sterilization were achieved.

Checks on the operation of the autoclave may also be carried out by means of spore strips. These strips contain the relatively heat-resistant spores of *Bacillus stearothermophilus* which should display no growth on incubation after they have undergone the sterilization process.

Temperature-sensitive autoclave tape is a useful aid for distinguishing between sterilized and non-sterilized material. However, these tapes usually show that the items reached the temperature of sterilization but do not indicate the time maintained. They are not checks of sterilization.

5.3.5. Recording

Every sterilization process must be recorded in writing. The following points must be covered:

(i)   date, time and serial number, if any;

(ii)   the chart record of the process;

(iii)   the target steam pressure and the recorded pressure;

(iv)   the temperature curve and temperature achieved;

(v)   the period of the overall sterilization process and that of actual sterilization;

(vi)   the inspector's initials.

These records are valuable checks on each batch processed.

5.3.6. Maintenance

Maintenance is essential to the correct functioning of autoclaves. Planned preventative maintenance can be carried out by an appropriately trained engineer in a documented programme. Periodic maintenance involving validation checks must be carried out by an accredited supplier and is usually part of a maintenance contract. The following items must be considered, depending on the type of autoclave:

(i)   valves for steam, process water, coolant and compressed air;

(ii)   steam filter;

(iii)   door packing;

(iv)   ventilation filter;

(v)   sterilization chamber;

(vi)   steam distributor;

(vii)   recorder;

(viii) vacuum leakage test;

(ix)   jacket pressure;

(x)   sterilizer chamber pressure;

(xi)   sterilization temperature of the chamber and separate sensor;

(xii) general points such as leakage and unusual noises.

5.3.7.   Safety

Detailed instructions must be available and displayed for correct, safe operation.

To prevent steam in the laboratory area, a condenser should be fitted in the blow-off pipe. If no condenser is fitted, the steam can be removed directly through the exhaust. In that case, the exhaust must be made of heat-resistant material.

5.3.8.   Decontamination

The autoclave should never be used for sterilizing media and decontaminating bacteriological waste in the same load. In order to achieve a sufficiently high temperature in the container for decontamination, an autoclave should be used preferably with a vacuum facility. A filter must then be included in the steam discharge pipe in order to prevent the distribution of aerosols into the laboratory area.

5.4.   Media preparator

A disadvantage of autoclaves with heating elements is the long time required for heating up and cooling down, which may detrimentally affect the quality of the nutrient base and may even produce growth- inhibiting substances by destruction of the components. The time can be reduced by using an autoclave with forced draft cooling. However, these are very expensive. A shorter processing time can also be achieved by resorting to a media preparator. The advantages of a media preparator are:

(i)   the nutrient base is dissolved during continuous stirring and heating;

(ii)   sterilization takes place immediately afterwards;

(iii)   distribution of the nutrient base can be automated;

(iv)   there is good process control, because the entire preparation is recorded.

The disadvantages are:

(i)   it is difficult to check the pH, which must be done afterwards;

(ii)   there is a lower limit for the quantity of medium to be prepared, generally 1 litre.

(iii)   only one medium type can be prepared per process.

5.4.1. Specifications and calibration

In principle, a media preparator is an autoclave and the apparatus must comply with the same specifications as stated in Section 5.3. Its calibration should follow the indications given in Section 5.3.1. However, because of its generally compact design, this is usually not entirely possible.

The essential requirements are:

(i)   sensitive and precise control, e.g. with the aid of a microprocessor, with time and temperature adjustable;

(ii)   recording and checking of the temperature;

(iii)   sterilization cycle temperature range from 70°C to at least 121°C, with an accuracy of 1°C;

(iv)   a temperature range of 35 to 80°C for distribution of the medium;

(v)   continuous stirring facility for the medium during preparation;

(vi)   no possible contamination by water from the water jacket around the medium vessel;

(vii)   a built-in bacteria filter to admit sterile air during cooling;

(viii) possibility for distribution of the sterile medium under the right conditions.

5.4.2. Calibration

A departure from the measured value by 1.5°C or less is permissible. Adjustment of the apparatus must be made by suppliers having calibration accreditation.

5.4.3.   *Validation*

Validation should be undertaken as described in Section 5.3.2, although this description is probably too detailed for a media preparator. However, those points applicable should be observed.

5.4.4.   *Use*

In a media preparator, the entire process takes place within the apparatus. The correct quantity of nutrient substances is added to the water in the preparation vessel, and then dissolved under continuous stirring and heating. Subsequent sterilization ensures a standard procedure. After preparation, the plate pouring should be checked to ensure an even surface

5.4.5.   *Checks*

Checks are made on the process by recording time and temperature.

5.4.6.   *Recording*

Most appliances are equipped with a recording system including a built-in chart recorder. In principle, the same points must be ensured as indicated in Section 5.3.5.

5.4.7.   *Maintenance*

Maintenance does not differ from the description given in Section 5.3.6, on the understanding that additional attention must be paid to the water in the water jacket (possible cause of contamination). Since all kinds of media are prepared in the preparation vessel, this part must be carefully cleaned after each use. It is recommended that a maintenance agreement is made with the supplier for the media preparator.

5.4.8. Safety

Apparatus of this kind is generally accompanied by manufacturer's instructions, both for the autoclave section and to guarantee sterility of the medium.

5.5. Pressure cooker

When facilities with proper autoclaves are not available, a pressure cooker may be used for sterilizing purposes. This apparatus is usually equipped with a valve, and instructions are given which permit operating at 110 or 121°C. However, it is necessary to regulate the pressure (temperature) on the basis of an accurate pressure gauge with a proper scale (see Section 5.2(b))

5.5.1. Calibration

The pressure gauge or temperature reading should be checked for accuracy of indications. A departure of the reading value by 1.5°C or less is acceptable.

5.5.2. Validation

Initial validation is important and periodic revalidation is advisable. Physical indicators (pressure gauge, thermocouple, thermometer) should be checked against those which have been calibrated.

5.5.3. Use

The pressure cooker may be used for quick sterilization of small volumes of media, keeping the following points in mind:

(i)   care should be taken for complete air removal;

(ii)   sterilized items must be removed as soon as possible to avoid excess wetting;

(iii)   loading should be with items of the same volume;

(iv)   care should be taken to keep the sterilization temperature within minimum fluctuation during sterilization time;

(v)   the water volume in the cooker should be sufficient for the whole sterilization cycle;

(vi)   care should be given for removal of steam from the laboratory area during the sterilization time.

This type of apparatus has the disadvantage that a relatively longer period of heating is required for a sterilization cycle than for autoclaves with a separate steam generator.

5.5.4.   Check

The efficiency of each operating cycle must be checked by biological indicators, e.g. *Bacillus stearothermophilus*.

5.5.5.   Recording

Records of the process cycle including load, air removal period, pressure achieved and sterilization time must be made.

5.5.6.   Maintenance

Maintenance must be regular and should include:

(i)   cleaning of the interior for removal of water deposits or media (it is advisable to fill with deionized or distilled water);

(ii)   inspection of valve, pressure gauge and thermometer fittings on the sealing cap;

(iii) points of leakage and unusual noise.

5.5.7.   Safety

Regular and special attention should be given to the fitting and safety points for the sealing cap.

5.5.8. *General remarks*

Well-controlled sterilization procedures have defined temperature/time exposure, warming/cooling times no longer than necessary and minimum frequency for performance checking of culture media. For quality control and safety, sterilization must be performed only by specially trained and experienced persons. Each sterilization cycle must be recorded in a special book and signed by the responsible person. Sterilized items must be marked with respective process numbers and dates and be well protected during storage.

5.6. Hot-air sterilizers (dry sterilization)

A high temperature (170–180°C) for 1–2 hours is necessary for dry sterilization. Hot air is circulated in a sealed cabinet (oven) by convection, conductance and radiation. To achieve a temperature of 175°C in the load, it may be necessary for the temperature on the dry sterilizer to be set rather higher (e.g. 180°C).

Materials that are heat resistant or those that do not allow or withstand steam penetration, e.g. oils or powders, can be sterilized by the hot-air process.

5.6.1. *Specifications*

The heating capacity of the sterilizer must allow the temperature to rise to 180°C in order for all parts of the loaded items to attain a temperature of 175°C. A fan facilitates the proper distribution of heat within the dry sterilizer. A clock with an alarm is a useful option for setting the appropriate time of at least one hour, excluding heating-up time. Also, a built-in temperature control may be used following the manufacturer's specifications. However, it is better to monitor and record the sterilization process with the aid of an appropriate sensor located in the largest item inside the chamber.

5.6.2. *Calibration*

The manufacturer supplies equipment with particular specifications that may not always meet the user's requirements. Therefore, the user will have to revalidate the apparatus before accepting it in his laboratory. If the apparatus does not meet user specifications it must be adapted by the manufacturer. Calibration can be undertaken in house with commercially available calibration thermocouples or calibration thermistors.

5.6.3. Use

In order to sterilize reliably, the following points must be taken into account and laid down in procedures:

(i)   At a steady temperature, micro-organisms die off more slowly in air with low rather than high relative humidity. To achieve a uniformly long sterilization time, the temperature in a hot air sterilizer must be higher than in a steam sterilizer.

(ii)   The thermal conductivity of air is much lower than that of steam, so a longer sterilization period is generally required.

(iii)   A uniform distribution of the hot air over the load is important. The distribution of air is assisted by a built-in fan.

(iv)   Packing, if any, of the material to be sterilized must be heat resistant. Special sterilization paper and laminate bags will suffice. Plastic (e.g. bottle tops) is not generally resistant to the dry heat sterilization temperature.

(v)   The position of the material in the sterilizer.

(vi)   The relatively lengthy heating-up time.

(vii)   Ensure that the sterilizer has cooled down before removing the load or use heat-resistant gloves.

5.6.4. Checks

Checks on the operation of the hot-air sterilizer are less simple than steam sterilizers because no commercially available biological product is available for this purpose. Temperature-sensitive autoclave tape that changes colour after 45 minutes at 165–170°C is commercially available . This tape can be placed in a reference flask for checking when the desired temperature is reached.

5.6.5. Recording

Recording of dry heat sterilization processes can be made by the use of chart records if fitted, or by connecting a thermocouple probe for temperature and time recording. Items sterilized should be labelled with the date and batch number. The records should be initialled.

5.6.6. Maintenance

Maintenance of the dry sterilizer is limited to cleaning the inside and the outside.

5.6.7. Safety

The following safety precautions should be noted by all users:

(i)   Open the door carefully if the sterilizer is still hot.

(ii)   Allow the material to cool off or use heat-resistant gloves in order to remove articles from the sterilizer. Articles should be labelled: "This material may be hot".

(iii)   On no account place hermetically sealed flasks in the dry sterilizer because of the risk of explosion.

5.7. Filter sterilization

In filter sterilization, the bacteria in a fluid can be trapped on a filter which has a pore size of 0.2–0.22 mm. These filters are known as bacteria or ultra filters and many ready-for-use varieties are commercially available. The technique is used for:

(i)   liquids that cannot withstand high temperatures as they may be chemically changed or inactivated;

(ii)   components in media which undergo undesirable reactions during heat sterilization with other substances.

The fluid to be sterilized is filtered through a suitable sterile diaphragm filter under vacuum. Pressure filtration is also possible. The use of filters mounted on injection nozzles is a simple and commonly used form of pressure filtration.

5.7.1. Specification

For specifications of filters and filter apparatus the reader should consult the various manufacturers' manuals. It should be verified whether they meet the in-house requirements and those necessary for the particular determination. The filter's pore size, the material, the diameter and the shape of the filter should be taken into account. Ready-made disposable filters are generally sufficient for filtering small quantities. They are suitable, for example, for use with injection syringes.

5.7.2. Use

Disposable filters sterilized by the manufacturer are used with disposable syringes for the sterilization of small volumes. Larger filtering devices must be sterile; this can be achieved by steam sterilization. In addition, it must be possible to remove the vacuum in the suction flask through a bacteriological filter in the vacuum hose. Complete sterile sets are also available.

The filter mounting and vacuum flask should be packed separately in laminate bags or sterilization paper. Sterilization of the whole apparatus, with the membrane filter already attached, is preferable.

5.7.3. Checks

The operation of the filter can be checked by testing the filtrate for sterility by determining the colony count (pour- or spread-plate method). Checks on the suction apparatus or pressure apparatus will depend on the system used. With vacuum pumps, vacuum control and regulation is important. Using a too strong vacuum can damage the filter. If compressed air is used, it must be clean and sterile. The pre-filter used will need to be checked for effective operation and should be regularly replaced.

5.7.4. Recording

The sterility of the filtrate should be verified periodically and should be recorded. The sterilization date should be entered on the filtrate container.

5.7.5. Maintenance

The vacuum or compressed air installation needs regular maintenance. In view of the wide variety of equipment and systems available no general guidelines can be laid down for this purpose. Manufacturers' recommendations should be followed.

5.7.6. Safety

In addition to the prescribed safety requirements for the vacuum and pressure installations, safety measures must be implemented at the workbench against implosion and explosion of suction apparatus.

5.8. Membrane filtration apparatus

Membrane filtration is a technique in which a known measured volume of a liquid, usually water, is drawn by vacuum through a membrane of sufficiently small pore size to trap all the micro-organisms in the sample.

5.8.1. Specifications

The apparatus consists of a manifold accommodating one or multiple filter funnels which are easily removable and a vacuum source connected by silicone tubing. The porous support for the membrane must provide sufficient support and prevent damage to the membrane.

5.8.2. Calibration

The filter funnels will require regular calibration of the volume held, as the markings do not always correspond to the precise volume. Calibration is carried out by pouring a measured or weighed volume into the funnel and correcting the markings as necessary.

5.8.3. Use

The problems that may arise in use relate to cross-contamination between samples and contamination of the whole apparatus. Cross-contamination during filtration is prevented by using loose-fitting lids on the funnels and is controlled between samples by disinfecting or sterilizing the removable filter funnels and lids between each analysis. Disinfection can be carried out by completely immersing them in a bath of boiling water for one minute. If contamination is suspected, the manifolds can be disinfected by passing boiling water through, taking precautions to avoid scalding. This equipment should always be drained after use by tipping away any residual water.

5.8.4. Checks

Blank samples will detect contamination of the apparatus or failure of disinfection. These samples should be filtered at the beginning, middle and end of each batch of split samples (see Chapter 8) and can be filtered through different pieces of apparatus as an additional check.

5.8.5. Recording

The identity of the check samples needs to be carefully recorded so that any actions required, due to failure can be easily taken.

5.8.6. Safety

The production of a vacuum involves a risk of implosion of the vessel used and requires the fitting of expanding plastic mesh over the vessel to contain broken glass. To prevent water passing into the vacuum pump, a 'vacushield' should be inserted in the tubing. The collection vessel will need to be emptied frequently.

5.9. Incubators

Incubators are cabinets with heating and/or cooling elements that aim to keep the required temperature in the cabinet within the tolerance limits of a given method.

Temperature *variation* means the difference at any time between two points in the incubator.

Temperature *fluctuation* means a short-term change in temperature at any point in the incubator during operation.

5.9.1. *Specifications*

Factors influencing the performance of incubators include:

(i) The heating element:

A strong heating element will restore the temperature quickly but will cause large fluctuations. A weak heating element will restore the temperature slowly but with only small fluctuations.

(ii) The thermostat:

To cope with small fluctuations in temperature a thermostat must have high sensitivity and a low differential value (e.g. 0.01°C ± 0.1).

(a) Incubators with forced air ventilation have a rapid heating-up/cooling-down time and a low temperature fluctuation.

(b) The overall temperature fluctuation depends on the specifications of the thermostat, the power and position of the heater, the presence or not of forced air ventilation and the size of incubator.

(c) The temperature fluctuation/variation of incubators quoted by suppliers apply generally to approximately 5°C above ambient temperature. It is therefore obvious that for incubation at ambient temperature or lower, an incubator with a cooling system must be used.

(d) To control the temperature fluctuation, it is advisable to record the temperature with a temperature recording system, possibly fitted with an alarm that is triggered when the temperature exceeds a fixed interval.

(e)   For pre-incubation at a lower temperature, incubators fitted with a time-switch temperature control of at least two levels can be used. The time required from switching over to the eventual incubation temperature should be reached preferably within two hours.

5.9.2. *Calibration*

For ordinary incubators at 22–42°C a combination of temperature variation and fluctuation must not exceed ±0.5°C. For incubators at 44°C, the overall fluctuation must not exceed ±0.25°C.

5.9.3. *Validation*

Before an incubator is put into service, it must be established that the temperature tolerance required is achieved in more than one position in the incubator. Calibrated thermometers immersed in 50 ml glass flasks containing 25 ml glycerol must be used to check the temperature inside the incubator. With small incubators, measurements should be made at three different points (centre, top, bottom) and with large incubators, e.g. above 300 litres, at six different places using temperature sensors, immersed in glycerol. The validation should be repeated under the maximum permitted load according to user instructions. The validation must also be repeated periodically and particularly after repairs, or if the incubator has been moved to another place/environment or if there are major changes in the laboratory temperature. The control thermometers/sensors must also be periodically calibrated.

5.9.4. *Use*

Culture media in tubes, flasks, Petri dishes, plastic bags or anaerobic jars are incubated at a defined temperature. Proper loading of the incubator is very important. Stacks of Petri dishes, for example, should not be more than four to five plates high, and large anaerobic jars or flasks may not attain the correct temperature within the desired time. Incubators with a ventilation system allow more rapid heating and are less subject to temperature variation. However, they do have two disadvantages: increased drying out of the media may occur and there is an increased risk of cross-contamination. Nevertheless, the advantages usually outweigh the disadvantages, which may be avoided, for example, by using non-vented Petri dishes or protecting the plates in plastic bags, etc.

5.9.5. Checks

The incubation temperature is a key parameter for the determination of microbial strains. Therefore, its proper functioning is of paramount importance. The incubator's temperature should be checked and recorded every morning at the end of the overnight incubation. Temperatures measured at other times will be affected by door opening and loading of the incubator. The temperature is measured using a thermometer held in water/glycerol to the correct immersion level on the in-use shelves.

The use (if economically possible for the laboratory) of a system of sensors connected to a central computer will keep a continuous temperature record of many incubators and of the temperature differences on the upper and lower shelves in the incubator.

5.9.6. Maintenance

Regular maintenance must be carried out. Cleaning and removal of media remnants should take place regularly and should be recorded in the incubator log.

5.9.7. Double jacket incubators

The jacket is filled with distilled or deionized water and the heating elements, usually of low power, warm up the water which maintains the desired temperature with the aid of a thermostat. The water jacket heats up the interior of the incubator from all sides and such incubators therefore show little variation/fluctuation in temperature.

5.9.8. Specifications, calibration, validation

See Sections 5.8.1, 5.8.2, 5.8.3, 5.8.4 and 5.8.5.

5.9.9. Maintenance

Care should be taken that the incubator is always filled with water according to the water indicator column and refilled if necessary.

5.10. Water baths

Water baths are used for incubation of media, for rapid pre-heating of media to be transferred in incubators at the selected temperature and for melting and/or maintaining agar media in the molten state.

5.10.1. Specifications

Water baths used for culturing or for heating up media at critical temperatures, usually above 37°C, must be able to control the temperature fluctuation within very narrow limits. The temperature control depends on the sensitivity and the differential value of the thermostat, the power of the heating unit and the efficiency of a water circulating system. A cover is recommended to prevent evaporation and to facilitate temperature control. Water baths for melting and/or maintaining agar media in the molten state are permitted to have a greater temperature fluctuation.

5.10.2. Calibration

Water baths for incubation must have a temperature fluctuation of only ±0.5°C at 22–42°C and ±0.25°C at 44°C. A fluctuation of ±1°C is acceptable for baths used for molten agar bases.

5.10.3. Validation

Each new water bath must be validated before use, and during use. Validation must also be repeated after repairs or after moving the apparatus to another place/environment.

5.10.4. Use

Water baths with an efficient control of temperature (±0.1°C) are used for incubation of culture media or for heating up media to the desired temperature after which the media are transferred to an incubator at the same temperature. Other water baths (±1°C) are used for melting and/or maintaining melted agar bases.

In comparison with incubators they have the same advantages, namely small temperature variations and fluctuations, in particular those equipped with a water circulation system. However, when Petri dishes are incubated in water baths, they must be in safe, watertight canisters or jars. Care must also be taken to control possible cross-contamination. The limited space in water baths in comparison to incubators restricts their use.

5.10.5. Checks

The actual temperature and the possible fluctuations in water baths must be checked with calibrated thermometers having a suitable scale (0.1 division). Temperature variations between different points within the same water bath must be controlled.

5.10.6. Recording

Temperature records must be kept for each water bath.

5.10.7. Maintenance

For each water bath, a maintenance cycle should be established and the service should be recorded. Removal of glass, remnants and media remains must be made regularly. The water bath is cleaned by heating to 80°C for one hour, and after allowing for cooling, the water is removed by siphon. The bath is cleaned with citric acid 0.3% or sorbic acid 0.1%, debris is removed and the bath refilled with distilled water and reset.

5.11. Microwaves

Microwave ovens have been used for melting down solidified nutrient bases. It should be noted that optimum radiation intensity, timing and quantity of nutrient base can only be fixed empirically. The major disadvantage of microwave ovens is that there are always cool spots in the chamber which will lead to local overheating or local insufficient heating and therefore differences in quality of media. The agar medium can be easily overcooked and hermetically sealed bottles may explode. It is advisable that the media melted in the microwave oven be checked for quality performance with those obtained by conventional techniques. The use of a microwave oven is not recommended for everyday use but only exceptionally, and full account of the possible problems should be taken.

5.12. Thermometers

Thermometers and thermocouples are used for temperature measurement and recording in microbiology. Liquid-in-glass thermometers may be of high accuracy, namely in the order of 0.02°C. Ordinary thermometers achieve a maximum error of usually ±1°C. Precision or reference thermometers with certificates may be bought from manufacturers at a high cost. Alternatively, thermometers with the desired scale division (e.g. 1, 0.1 or 0.02°C) can be calibrated by other metrological institutes or in the laboratory by careful comparison with reference thermometers. This can be done by taking three readings at each of three temperatures at and around the in-use temperature. Any consistent temperature deviation of the test thermometer from the reference thermometer must be applied as a correction factor in all subsequent measurements made with the former. Recalibration of thermometers should be carried out every five years or more frequently if the checks at the reference point indicate a significant change (e.g. >0.1°C). For thermometer specifications, ISO standards should be consulted. (NEN-ISO 386, NEN-ISO 654 and NEN-21770 or NAMAS-N1S7-June 1991).

5.13. pH meters

The performance of culture media is often influenced by their pH. Determination of the pH is an important quality aspect of media production and quality control. pH measurements can be made on liquid media or reagents and/or solid agar bases.

5.13.1. Specifications

A pH meter should preferably be fitted with a built-in temperature compensation system and with two regulating facilities for adjustment. The reading scale must be graded to 0.01 unit of pH and the reproducibility of 0.01 units of pH. The electrodes must be suitable for measurement of pH at temperatures from 0 to 80°C. A combined electrode equipped with a flat head can be used to measure the pH of solid agar media.

5.13.2. Calibration

The pH meter must be calibrated at least once every day by using freshly prepared buffer solutions. Two buffer solutions of different pH values (usually of pH 4.0 and 7.0) are used. A calibrated mercury thermometer with a scale division 0.1°C and a maximum error of 0.1°C is used for the temperature measurement. The instructions for the apparatus must be followed for adjustment before each use. The adjustment of pH meters is dependent on the surrounding temperature and the temperature of the materials that are to be tested.

5.13.3. Checks

The electrodes are subject to ageing especially when solid media are used. This becomes apparent when there is a slow response during calibration. The final pH signal must be apparent within one minute with a stable response (±0.03 pH units) and sensitivity better that 95%. The sensitivity of electrodes can be checked by measuring the difference in mV potential between pH 4.00 and pH 7.00. The difference should be 172–171 mV. If the difference is less than 172 mV but higher than 150 mV, the electrodes should be regenerated. If the difference is less than 150 mV, the electrodes should be discarded. When electrodes do not perform properly, the troubleshooting chart that comes with the electrode should be consulted.

5.13.4. Use

pH meters can be used on liquid media, reagents or solid agar media. A sample of the liquid medium is transferred to a beaker, preferably after stirring on a magnetic stirring device. The temperature should be read and the pH meter adjusted to the measured temperature. The electrode should be placed into the liquid medium until the signal is stable.

The solidified agar medium from a Petri dish is cut into pieces and transferred to a test tube, to half fill the tube. About 3 ml distilled water are added and after a minimum of five minutes, the pH is measured. It is essential that the electrode is properly cleaned after use, e.g. by rinsing with hot tap water at 80°C to remove the agar, unless the manufacturer recommends a different procedure.

5.13.5. Maintenance

Proper maintenance of the instrument is important. The glass electrodes should be cleaned at least once a week or more frequently when necessary. For electrode cleaning, a pepsin solution (commercially available) can be used. Fat can be removed by acetone rinsing, but prolonged immersion of electrodes in acetone may cause desiccation of the membrane and will require equilibration in water. The manufacturer's instructions regarding storage and maintenance should be followed. Every procedure for maintenance or cleaning must be recorded and signed in a special book.

5.14. Balances

Balances may be of different types and weighing capacities. Many laboratory balances have two pans with three knife edges. Two of the knife edges support the pans.

Single-pan balances may be of the ordinary analytical type or of the electronic type.

Whatever the type of balance it performs best when clean and operating in an environment that is free from vibration, dust or rapid environmental change. Any tilting of the apparatus should be strictly avoided.

5.14.1. Calibration

Calibrated weights must be used daily to check the accuracy of electronic balances. Also the zero and tare mechanism must be checked prior to loading and after unloading. Repeatability is checked by 10 repeated weighings with not less than four weights within the weighing range (1/4, 1/2, 3/4 and full). Adjustment of zero and tare zero should be made according to manufacturer's instructions. Weights can be calibrated at accredited laboratories or institutes. For two-pan balances, the zero adjustment and the use of calibrated weights is also important. The accuracy should be 1% or better. Some manufacturers are accredited for recalibration of their apparatus and may include it in their maintenance contract.

Frequency of calibration depends on the type of balance, the accuracy required and the specifications of the manufacturer. A full calibration once a year is recommended.

5.14.2. Checks

Checks of zero set and tare zero (if fitted) must be on a daily basis or before use. For electronic balances, it is advisable to check possible disturbances and errors that may be caused by electrical interference.

5.14.3. Maintenance

Balances should be always clean. Weights should be cleaned regularly and must be protected from dust or dirt. The cleaning procedure should be done at a regular time, recorded and signed in a special book.

5.15. Deionizing and distillation apparatus

For the preparation of culture media, the water must be free from heavy metals (especially copper) and other substances which can affect bacterial multiplication. This can be achieved by distillation, demineralization, reverse osmosis, or a combination of these purification procedures (Fossum, 1982). The conductivity of freshly produced water from a deionizing or distillating apparatus must be monitored frequently, e.g. every week. An efficient installation should deliver water with a conductivity of less than 2–4 μS. If higher conductivity values are found, the deionizing resins must be regenerated or replaced. A disadvantage may be the presence in the effluent of breakdown products from the ion exchange resins.

A viable count at 22°C of 1000/ml or more micro-organisms in the water is unacceptable. Multiplication of microbes (contamination) in stored distilled/deionized water can be avoided by cleaning the storage tank regularly or by storage of the purified water at 80°C. The distillation apparatus should be cleaned regularly and especially when deposits begin to appear. The same applies for storage containers. Every cleaning, replacement, and maintenance procedure should be recorded and signed for in a special book.

5.16. Cold storage

Refrigerators or refrigerated rooms (cabinets) are used for two main reasons:

(i)   to store samples and chemicals;

(ii)   to store prepared media or their perishable supplements and ingredients.

The specifications for each category of storage facility depends on the type of items to be stored. For storing samples or chemicals, a temperature range of 0–4°C with a fluctuation of ±1°C is acceptable. For storing media or for cold rooms, 4–6°C, with a fluctuation of ±2°C is acceptable.

In cold rooms, an efficient air circulating system helps to maintain continuous and uniform refrigeration of the stored items. With prolonged storage, media may suffer desiccation.

The accuracy of the temperature should be verified with thermometers, electronic thermometers or continuous temperature recording devices. All should be calibrated against precision thermometers at regular intervals. Possible temperature variations between different positions, especially in cold rooms, should also be checked. The refrigerator/cabinet temperature should be recorded daily in a special book or by means of a temperature recording system.

Cleanliness should be maintained by removing leftover media or liquids and, depending on the type of refrigerator, by defrosting at regular intervals. Washing with a sodium bicarbonate solution may help to maintain refrigerators free from fungi.

The temperatures of freezers at –20°C and possibly at –40°C and –70°C or when liquid nitrogen is used for storage of cultures or supplements/ingredients should be checked daily. The temperatures should be recorded in a special book. There should be an alarm system and a planned response to the alarm when fluctuations of temperature are above the set limits.

As a general principle, media, reagents, test samples, and biological materials must be stored in separate designated cold-storage areas.

5.17. Anaerobic incubation

Anaerobic incubation can be carried out in jars specially designed for the purpose or, if large numbers of plates are to be handled, in an anaerobic cabinet. In anaerobic jars, the anaerobic atmosphere is produced by the addition of hydrogen to the jar containing a platinum catalyst, in the presence of which the oxygen reacts with the hydrogen/ carbon dioxide/nitrogen gas mixture to produce water. Alternatively, the hydrogen may be produced by a gas generation sachet to which water is added. The jar is then placed in an incubator to achieve the correct incubation temperature. Anaerobic cabinets are really large anaerobic glove boxes which allow the operator to work continuously.

5.17.1. Checks

The anaerobic atmosphere is verified by the growth of a strict anaerobe such as *Bacteroides melaninogenicus* and the failure of growth of a strict aerobe such as *Pseudomonas aeruginosa*. Subcultures of these organisms should be included on each occasion a jar is used, or daily in the case of the anaerobic cabinet. Methylene blue indicator strips, which turn colourless in anaerobic conditions, may also be used.

5.18. Colony counters

This is an apparatus with a transparent screen and illumination, equipped with wide, but rather low- magnification lenses for counting colonies on plates. An electrical or manual counter device helps with recording the colony count, which is usually done by marking the colonies (on the Petri dishes) to avoid double counting of the same colonies.

For automatic colony counters, frequent checks for accuracy are necessary; Petri dishes with particles embedded in plastic can be used. The size of the particles should mimic those of colonies in agar, and several plates with differing numbers should be used.

5.19. Microscopes

Ordinary light microscopes are equipped with three objective lenses (5×/10×, 40×, and 90/100×), and two ocular (5×, 10×) lenses. Many other types of microscope are used for particular microbiological examinations and may be equipped with a special type of illumination, e.g. phase contrast, ultraviolet or epifluorescence. Calibration of fields can be done by using specially calibrated microslides for counting/estimation of microbes fixed on slides. A check on the power of UV illumination produced is important in fluorescent microscopy. For all types of microscope, cleaning after use and regular maintenance service is a task for which designated members of staff are responsible. Maintenance service should be recorded and signed for in a special book.

5.20. ELISA plate readers and plate washers

The enzyme-linked immunosorbent assay (ELISA) is increasingly used for the semi-automated detection of *Salmonella* and *Listeria* in food. Labelled antibodies are utilized to detect the target organisms in the sample. The test is carried out in strips in 96-well microtitre trays and washing and reading by automated pieces of equipment.

5.20.1. Specification

The plate readers and washers should be set up by the manufacturer on installation and each piece of equipment will require a 12-monthly maintenance check.

5.20.2. Checks

The daily checks involve a visual examination of each plate before loading it onto the plate reader to look for splashes, pieces of debris in the wells and empty wells. This will ensure that artefacts are detected and do not influence the result produced by the plate reader.

Internal controls and blanks must be included in each batch of analyses.

The plate washer should be checked daily; this involves a visual examination to ensure that each of the channels is delivering the wash material.

5.20.3. Safety

There should be a disinfection policy for these pieces of equipment. Protection of the vacuum system of the plate washer is necessary.

5.21. List of equipment and technical specifications

1.   Horizontal or vertical autoclaves. Pressure regulation precision 0.05 Bar. Temperature fluctuation ±1°C, at 115°C and 121°C.

2. Hot-air sterilizers reaching 180°C, maintaining this temperature for one hour.

3. Incubators
(a) for 25–50°C thermostatically controlled, vented/non-vented, precision ±0.5°C,
(b) for 44°C, vented/non-vented, precision ±0.25°C.

4. Waterbaths:
(a) for 20–70°C. precision 1°C,
(b) for 44°C with water circulation unit, precision ±0.25°C.

5. Thermometers calibrated/mercury or with coloured fluid:
(a) scale division, 0.1°C, maximum error 0.1°C
(b) scale division 1°C.

6. Temperature registration system, sensors, computer and recorder.

7. Anaerobic jars or anaerobic cabinet (glove box).

8. Deionizer/distillation apparatus, water conductivity <2 μS (<3 μS).

9. Conductivity meter. Division 1 μS or better 0.5 μS.

10. Balances:
(a) laboratory/electronic, precision 1 mg,
(b) laboratory/electronic, precision 1g.

11. pH meter: scale division 0.1 pH units. Temperature adjusting facility.

12. Homogenization apparatus:
(a) blender >8000 c/m or with variable speed,
(b) stomacher of appropriate volume capacity.

13. Manual colony counter. Automatic counter after calibration.

14. Media dispenser volumetric and optional automated media preparation — plate pourers, automated diluters.

15. Refrigerators and freezers. Refrigerators thermostatically controlled at 0–6°C or rooms (±1°C). Freezers –20°C, and optional, –40°C, –70°C or liquid nitrogen.

16. Filtration units for membranes 47 mm diameter, withstanding flaming or boiling, filtration units for sterilizing media.

17. Safety cabinets, laminar flow.

18. Microscope with at least three objective lenses. Calibration of fields (optional).

19. ELISA (plate washers, plate readers). Calibration on installation.

Chapter 6

Materials

6.1. Introduction

Major items of equipment have been discussed in Chapter 5. There are, however, numerous small items and materials used in food and water microbiology laboratories (Table 6.1). They cannot be taken for granted; their specifications are important and lack of control may lead to erroneous results which may not be apparent. This chapter discusses these small items of equipment and materials, the possible effects that lack of quality control may produce, and suggests

Table 6.1

Materials used in food and water microbiology laboratories

Sample containers	Membrane filtration
water	membranes
food	pads
Glassware	Deionized/distilled water
Pipettes	Chemicals
Petri dishes	Culture media
Metal utensils	Reference cultures
Dilution tubes	

the measures that need to be included in a quality assurance programme. It will probably be impossible to introduce all these control procedures into all laboratories in one step; instead, laboratories should rely on end product testing and internal quality control initially and resort to these recommendations when investigation is required. They will then of course become part of the quality assurance programme.

There are many check procedures, and guidance is given for all the materials listed in Table 6.1. Often the manufacturer can be asked to provide materials to a documented specification and the checks become minimal. In the control of bacteriological culture media which are all important in the measurement process there is a balance between process controls and end product controls. It will vary from laboratory to laboratory for the measurements being made and whether resampling can be carried out.

6.2.   Sample bottles and containers

Sample bottles are used to transport the sample from the sampling point to the point of analysis; they must be fit for the purpose such that the sample analysed truly represents the sample taken and the micro-organisms present. The sample bottles or containers must therefore be sterile, non-toxic and made of suitable material.

6.2.1.   *Sample bottle specifications*

Water testing

The sample bottle must be of an appropriate volume for the analyses requested and any further analyses that may be required.

coliforms	100 ml
faecal coliforms/*E. coli*	100 ml
faecal streptococci/enterococci	100 ml
sulphite reducing *clostridia*	20 ml
colony counts	4 ml
staphylococci	100 ml
Pseudomonas aeruginosa	100 ml

Air space is necessary in the bottle to allow thorough mixing before processing, therefore 500 ml bottles will be sufficient for the majority of analyses. In some cases, a smaller volume may be sufficient, for example, for simple surveys in bathing areas the measurement of *E. coli* + *enterococci* by microtitre plate methods requires only about 10 ml, or in waste water even smaller volumes may be used for dilutions. There are advantages in using only one kind of sample bottle; stock management and cleaning is simplified and confusion over the correct sample bottle size is avoided.

In other situations, a 500 ml bottle will not be sufficient and larger volumes will be necessary. Mineral waters, bottled waters, water for haemodialysis, sterile water in hospitals or electronic factories and samples for the examination of individual pathogens will require a sample volume of 1000 ml. Examination of larger volumes in the measurement of viruses or *legionella* will require 5- or 10-litre containers.

For the detection of amoebae or *Giardia*, where up to 100 litres are examined, a concentration step is usually made on site using a cartridge filter which is then transported to the laboratory.

Food analysis

Food samples can often be taken in their original packaging. For sampling of non-packaged foods or sampling from packages too big for delivery to the laboratory, bottles (for liquids) or plastic bags (for solid foods) can be used. The plastic bags will be especially suitable if homogenization of the food for the resuspension of the bacteria is to be used.

Alternatively, standard containers such as plastic or glass jars capable of holding 200 g portions can be used.

6.2.2. Glass or plastic bottles

Wide-mouth sample bottles should be used for ease of sampling and to avoid external contamination. They can be made of low-alkali borosilicate glass or other non-corrosive glass, or of different polymer materials such as:

(i)   polypropylene (PP): semi-rigid, translucent, autoclavable;

(ii)   high density polyethylene (PE): semi-rigid, translucent;

(iii)   low density (high pressure) PE: semi-rigid, transparent;

(iv)   polystyrene: rigid, breakable, transparent;

(v)   polycarbonate: rigid, breakable, transparent.

6.2.3. Closures

The closure can be a ground glass or plastic stopper for glass bottles, or a plastic or metal screw cap in any case, or a plastic press-on lid attached to a plastic bottle or jar. Glass stoppers must be autoclaved separately from the bottle, or with a piece of paper between them, to prevent the stopper sticking. Metal caps, especially aluminium caps, can generate some toxicity when autoclaved and should incorporate a heat-resistant leak-proof liner. Bakelite and other materials can also give toxic by-products when heat sterilized, even in a dry oven, or induce pH changes. Some brands of cotton used to make plugs for glassware may become toxic if they are heated for too long at a too high temperature. Press-on plastic lids linked to the bottle or jar have several advantages in that they are almost as leak-proof as screw-caps and the lids can stand open, which facilitates filling and pipetting. When open, the lid remains linked to the bottle, so interchanges between bottles and closures are avoided and in addition, the lid is protected from contamination.

6.2.4. Plastic bags

Some plastic bags are available which contain thiosulphate to neutralize the residual chlorine of disinfected waters. The use of plastic bags for sampling water requires experience; usually, rigid or semirigid bottles are preferred. Plastic bags are suitable for the sampling of food.

6.2.5. Neutralization of disinfectants

Assessing the microbiological quality of water disinfected by an oxidant (chlorine, chloramine, bromine or ozone) requires the action of the oxidant to be stopped as soon as the sample is taken. This is usually achieved by adding a reducing agent such as sodium thiosulphate to the sample bottles. It has been claimed that *Legionella* are sensitive to sodium and that potassium thiosulphate is preferable, but with the concentration used for neutralizing normal chlorine concentrations no adverse effect of sodium has been detected.

Thiosulphate is not destroyed by autoclaving, but it could be by other treatments. If this is the case, it must be added aseptically as a solution sterilized by filtration, after sterilization of the bottles. Alternatively, it can be added in excess, in order to give a residue after sterilization of about 18 mg thiosulphate per litre. The theoretical concentration of sodium thiosulphate necessary to neutralize 1 mg of chlorine is 7.1 mg. Thus, 0.1 ml of a 1.8% m/V solution of sodium thiosulphate should be added for each 100 ml of bottle capacity.

Other products such as chelating agents have been recommended to protect bacteria from the toxic action of heavy metals such as copper or zinc. Ethylene diamino tetra acetate (EDTA) or nitrilo tri acetate (NTA) ($Na_3C_6H_6NO_6$) can be used as a filter-sterilized solution at a final concentration of about 50 mg per litre but should only be provided when necessary.

The neutralizing solutions, if added in small volumes (0.5 ml) may evaporate after sterilization and produce a ring precipitate round the walls of the bottle. In addition, if the bottle is rinsed before sampling the neutralizing agent will be eliminated. To prove that this has not been done, or that no confusion exists between bottles (e.g., cap for bacteriology on the bottle for chemistry and *vice versa*), the laboratory can use a tracer such as lithium chloride in the thiosulphate solution.

6.2.6. Sterilization of bottles

The outer surface of the bottles is always subject to multiple contamination events such as dust in the storage room, technicians' hands, contact with outer containers or contact with other flasks already dipped in various kinds of water. In these situations externally contaminated bottles may interfere with the quality of water when sampled by immersion. It is therefore useful to produce some bottles sterile both inside and outside and protected by outer bags. This will require sterilization with gamma rays or by ethylene oxide. The bag can then be opened just before sampling and can also serve as a glove to hold the bottle to provide maximum asepsis before being placed on a pole or other sterilizable sampling apparatus.

Autoclave sterilization of bottles

Autoclaving (121°C, 15 minutes) in moist heat is convenient but requires a loose closure, to allow the steam to replace all the air during the temperature rise, and to prevent plastic bottles from collapsing when cooling. The screw caps must be tightened after sterilization.

Dry oven sterilization of bottles

Heating in a dry oven for 2 hours at 180°C is convenient for empty glassware. Ground glass stoppers should be separated from the neck by a paper tape or a piece of string to avoid blocking during cooling.

Ethylene oxide sterilization of bottles

Polyethylene bottles can be sterilized by exposure to ethylene oxide gas but because of its toxicity the procedure has to be carried out in specialized workshops and time has to be allowed for the desorption of the ethylene oxide. It is therefore not used as a routine laboratory procedure.

Gamma rays

Exposure to gamma rays produced by a ^{60}Co or ^{137}Cs source or to β rays of sufficient energy (1 to 2×10^5 Gy) are very efficient sterilization techniques. They are available only in specialized workshops . There is no residual antibacterial activity but the material may be altered by polymerization after repeated irradiation.

6.2.7. Quality control of sample bottles

Sterility

The laboratory must ensure the sterility of the sample bottles, whether they are home-prepared or manufactured, whether they are made of glass or of plastic. A check for sterility must be a condition for acceptance of the delivery. A sterility test must also be made on the batch in use, at the rate of usually 1 per 100. This test relates to the batch of bottles after labelling, addition of neutralizing agents and preliminary storage.

The 'roll bottle' method is used, which consists of introducing 20 or 50 ml of melted nutritive agar (plate count agar) into the bottle to be tested and of lining the walls with agar by rotating the flask while cooling (under a trickle of water if necessary). Incubation at 20 to 22°C for five days should not reveal any growth of colonies.

The sterility of sample bottles can also be checked by placing 20 to 50 ml of thioglycollate broth inside, by rolling the bottle to wet the walls and incubating at 30°C for five days.

If the sterility of bottles can be guaranteed by process controls of the sterilization, these end product controls are not necessary.

Testing for the presence of neutralizing agents

The presence of thiosulphate may be checked by an iodometric method:

$$I_2 + 2S_2O_3^{2-} \longrightarrow 2I^- + S_4O_6^{2-}$$

Add 10 ml distilled water into the bottle and titrate with a 0.10 N iodine solution, using starch or thiophene as an end point titration agent.

Testing for residual toxicity in sample bottles

Residual toxicity in sample containers may arise from the washing procedure of the glassware, from the release of components or additives from plastic bottles and also from the sterilization process. If home-treated glass bottles are used, they can be checked for absence of residual toxicity with the rest of the glassware, at the same frequency. For disposable containers, especially plastics, each new batch should be tested.

Cultures of *Enterobacter aerogenes* are made in liquid media; one medium is made with boiled purified reference water, the other with boiled reference water previously in contact for 24 hours with the container to be tested. A difference greater than 15 to 20% between the numbers of colonies after 24 hours in the test culture and in the control gives evidence of toxicity.

The detailed procedure is described by Geldreich and Clark (1965).

6.3. Laboratory glassware

6.3.1. Specifications

Laboratory 'glassware' nowadays consists of glass or plastics. Glass containers can be made of:

(i)   soft glass (sodalime glass);

(ii)   borosilicate glass;

(iii)   heat-resistant borosilicate (Pyrex®, Duran®, etc.).

Borosilicate glass is preferred. Soft-glass items should not be used because they release alkali and change the pH of the media. Some are lined with a polyphosphate film and can be used for some purposes, once only. They cannot be re-used because the protecting coat will be destroyed by autoclaving. Volumetric glassware must be of class A and must withstand repeated sterilization without significant change in volume. Disposable glass pipettes should not be re-used. All glassware must be kept clean, free of residues of culture media. Metal with crevices, chopped or cracked glass and frosted or scratched plastic must be discarded.

Plastics are for single use only and must be discarded after that use.

6.3.2. Cleaning procedures

Glassware for microbiology should not be mixed with glassware for chemistry and should never be treated with sulphochromic acid. Whether done in a washing machine or by hand, the cleaning procedure must include a sequence with detergent at about 70°C, a rinse with clean (soft) water at about 80°C, a final rinse in distilled water or equivalent and drying. The washed glassware should be sparkling clean, free from acidity, alkalinity and toxic residues. Contaminated (culture) vessels should be autoclaved prior to cleaning. New glassware items should be soaked in distilled water overnight before use, and this soaking should be repeated if the pH is not neutral.

6.3.3. Quality control

Tests for pH, sterility and residual toxicity may be required as described for sample bottles.

6.4. Pipettes

6.4.1. Glass pipettes and disposable plastic pipettes

Bacteriological pipettes are of the TD type ('to deliver'), i.e. the nominal volume is contained between the graduation and the tip. If the tip is broken, the pipette should be discarded. These pipettes are plugged to prevent cross-contamination when pipette bulbs are used.

6.4.2. Automatic pipettes and tips

Fixed or adjustable volumes can be delivered by automatic pipettes provided with plastic conic tips where the liquid is aspirated by moving an air piston. Some can be provided with several tips (up to eight). The tips must be chosen so as to match the pipetting device without leak, and their shape must allow pipetting in narrow-neck tubes. Plugged tips may be required to prevent contamination of the piston.

6.4.3.   *Accuracy*

No graduation exists on the tips of automatic pipettes and the accuracy of the volume delivered relies only on the type of pipette for fixed volume models or on the setting of the adjustable piston. These pipettes must therefore be submitted to regular checks for accuracy. This is simply achieved by weighing with a calibrated precision balance the volume or an integer number of the volumes delivered, using water at 20°C.

6.4.4.   *Sterilization*

Plastic disposable pipettes are normally purchased in sterilized plastic bags, individually or by 10 or 25. The bag must be closed after withdrawing a pipette. Glass pipettes are generally sterilized in glass, aluminium or steel containers, in an autoclave or a dry oven. A pad in the bottom will protect tips from breakage. Tips of automatic pipettes are disposable, but it is necessary to clean the pipette at regular intervals or if the liquid comes into contact with the pipette. This can be done with a cotton swab impregnated with 70% alcohol.

6.4.5.   *Quality control*

Accuracy of the pipettes must be checked as indicated under Section 6.4.3. Sterility of the pipettes can be checked by pipetting sterile water and inoculating it in Petri dishes with melted agar. After hardening, incubation for five days at 30°C should be carried out. No colony should appear. Alternatively, a thioglycollate broth should be inoculated and incubated for five days at 30°C. No culture should appear.

6.5.   Petri dishes

6.5.1.   *Capacity and specifications*

Petri dishes are normally 90 or 100 mm in diameter, but some laboratories use smaller dishes for the incubation of membrane filters (diameter ±55 mm). The lids of Petri dishes are normally provided with three or four stops to allow aeration.

6.5.2. Sterilization and quality control

The same rules as for sample bottle should be followed.

6.6. Metal utensils

6.6.1. Wire loops, needles

Nickel-chromium wire is generally needed for making inoculation loops and needles. These must be cleaned with emery-cloth when oxidized. Platinum wire will not oxidize as quickly as Ni-Cr, and also cools more quickly.

6.6.2. Forceps and spatulas

Forceps and spatulas must be made of metal not subject to corrosion, and those used to handle filter membranes should have smooth ends to avoid damaging the fragile cellulose membranes.

6.7. Tubes and closures

6.7.1. Dilution tubes (bottles)

Microbiological examination of contaminated food and water requires dilutions, usually 1:10. These are obtained by transferring 1 ml of the sample into 9 ml of sterile water or diluent in a dilution tube. Larger volumes are sometimes used (e.g. 90 or 99 ml), in dilution bottles.

These dilution tubes or bottles are generally prepared in advance, so the exact volume must be dispensed in each item but also remain stable upon storage. The dilution tubes or bottles must allow a vigorous mixing. Screw-caps therefore are preferred.

In addition to quality controls applicable to all glassware, e.g. sterility, neutrality and non-toxicity (see above), a check of the diluent volume must be made, either on each batch after autoclaving or preferably at regular (e.g. monthly) intervals on those in use, to assess the effects of storage. A volume shift of 2% is an alarm level, and 5% a rejection level. The measurement of volume by weighing is recommended to check the dispensing procedure.

6.7.2. Cultures tubes and other vials, caps and plugs

Contrary to dilution tubes, culture tubes can be closed with cotton wool or loose plastic or plastic-lined caps, if their storage period is short. Caps cover the lid and upper part of the tubes. On the contrary, cotton (or paper) plugs do not and the tubes must be flamed when opened. Like paper, cotton no more longer forms a barrier to bacterial contamination if it is wet. Quality controls are as for other glassware items.

6.8. Membrane filters

6.8.1. Physical, chemical and biological specifications

Membrane filters (MFs) are an important component of bacteriological analysis and it has been shown that variations of quality can occur not only between brands but also between batches from one manufacturer.

MFs used for culture of bacteria are made of a reticulum of cellulose esters (acetate or a mixture of nitrate and acetate). Their porosity is equivalent to 0.45 μm.

It is not necessary to use MFs of 0.22 μm which are used for sterilizing liquids, as their flow rate is lower and they recover only a negligible proportion of minicells that have been forced to pass through 0.45 μm. New, wettable, autoclavable MFs made of polyether sulfone could be used in place of cellulose MFs.

MFs made of polycarbonate are flat films perforated with well-calibrated cylindrical holes. They are used for microscopic examination by epifluorescence and to concentrate some bacteria (e.g. *Legionella*) before re-suspension and culture in or on selective agar. Their flow rate is very low, and the transfer of nutrients from a nutritive agar to a colony through such an MF is not adequate for the enumeration of colonies on agar or pads.

Nylon MFs can be used instead of polycarbonate MFs for concentrating *Legionellae* before plating onto agar. The standard diameter for MFs is 47 mm. They are provided in paper bags, either individually or by 10 or 25. Some have to be sterilized, by autoclaving. Others are industrially sterilized.

MFs should not alter significantly the number of colonies to be counted, compared to other methods of seeding, such as incorporation into nutrient agar, or spreading onto agar. This supposes that the MFs retain all the microbial cells (or spores), that they are non-toxic and that they allow nutrients, selective substances and water to reach the growing colonies.

Routine quality control of MFs should include sterility testing and recovery capacity. If problems arise in the latter test, investigation of retention and toxicity should be carried out. MFs should not be re-used.

6.8.2. Quality control for sterility of membrane filters

Checks for sterility should be made on each new batch in use in the laboratory simply by seating several MFs ($\log_{10}n$) on nutrient non-selective agar, and incubating for three weeks at 30°C. This control is completed by 'blank' analysis made at least daily and should be repeated if the blanks reveal colonies.

6.8.3. Testing procedures

A global test of recovery should be performed on $\log_{10}n$ MFs of each new batch and repeated on a bigger number of MFs if unacceptable results are recorded with reference or standard materials in the daily control of membrane filtration procedures. This global test covers the retention, wettability, non-toxicity and even to some extent the sterility. It is fully described in ISO 7704-1985.

The principle is to seed at least five replicate samples (water or diluted culture) in parallel by membrane filtration and by surface plating on or incorporation in agar. For some selective media, used for specific micro-organisms, the formula has been adapted to MF procedure and differs from the formula for surface plating or incorporation. The reference for the test should therefore be carefully chosen, and a test must be made for each selective medium in use in the laboratory. With a non-selective medium, the count by membrane filtration should not lead to a loss greater than 20%. Details for the preparation of a test sample is given in Appendix B.

Alternatively, reference materials can be used to test the acceptability of a new batch of filters and pads. Details are given in Section 8.3.3. An example of a calculation is given in Appendix A.

When problems arise with membrane filters it may be necessary to perform further tests such as retention and total specific extractables as part of the investigation.

6.8.4. Retention

A diluted culture of about 10^3 cells/100 ml of a small (0.2–0.3 μm) bacterium, *Serratia marcescens*, is filtered through the MF under test, the filtrate is added to a nutrient broth and incubated at 30°C for 48 hours. Five replicates must be made, along with five blanks using distilled water instead of the *Serratia* suspension. The test is satisfactory if all broths remain sterile.

6.8.5. Total extractables

Five MFs are dried at 70°C for 1 hour, then cooled in a desiccator and weighed individually with a 0.1 mg precision balance. They are then extracted for 30 minutes in boiling distilled water, and re-dried, cooled and weighed as above. The amount of total extractables must not vary from batch to batch and must be equal to the specifications on the package.

6.8.6. Specific extractables

Ten MFs are soaked in 100 ml distilled water for 24 hours at room temperature. Then the lixiviate can be analysed for total organic carbon, NH_3- nitrogen conductivity, ions and metals.

6.9. Pads

6.9.1. Physical, chemical and biological specifications

Instead of an agar layer to support the MF for incubation, an absorbent paper pad can be used. The pad absorbs the liquid nutrient broth and should not bring any extractables or toxics to this medium.

6.9.2. Testing procedures

Absorption capacity

The absorption capacity may vary between different brands and it is important that the correct amount of culture medium is used. This should be determined before first use and checked whenever changes occur.

Toxic residues

Ten pads are soaked in 100 ml distilled water for 24 hours at 30°C and the lixiviate tested for pH, individual contaminants or toxic materials.

6.10. Deionized/distilled water

Water is one of the most critical factors in the preparation of microbiological media and reagents.

6.10.1. Apparatus

Deionized water from a reverse osmosis unit can be used for the preparation of media. If it is well operated, this apparatus will produce even better quality water than distillators. The unit must include one or two stages of deionizing ion exchangers, and, if necessary, to preserve the exchange capacity, especially where the water is hard, a reverse osmosis unit. The latter must be protected from chlorine by a charcoal filter. A 0.4 μm membrane is sometimes added. The conductivity must be measured daily or continuously on line and the ion exchangers replaced or regenerated if the conductivity rises above a set value (e.g. 1 or 0.1 μS).

Distilled water can be obtained from quartz, Pyrex® or stainless-steel apparatus. All connections should be made of the same materials, or of high quality plastics, such as PTFE, PVDF, polypropylene and polyethylene but not PVC. In hard water areas it is advisable to use deionized water obtained from a well-maintained ion-exchanger to feed the distillation apparatus.

6.10.2. Specifications

The conductivity should be less than:

>0.5 μS   with deionizer with or without reverse osmosis,
>0.5 μS   with double distillation on Pyrex® or Inox,
>0.3 μS   with double distillation on quartz,
1 to 3 μS   with single distillation on Pyrex®.

These values should be measured with a conductivity meter equipped with a sensitive cell (K < 1 cm^{-1}) immediately after production of the water, because exposure to the air and carbon dioxide absorption will result in a rise from 0.5 to 5 μS in 15 minutes.

In reverse osmosis, the conductivity may change if it is not properly maintained. Monodistilled water is also subject to flaws; it can become contaminated by the components of tap water such as ammonia, chlorine or chloramines and fluorides.

Deionized and distilled water should be kept in high-quality materials or in epoxy-fibre glass tanks, and be protected from dust, laboratory fumes (NH_2, HCl) and cleaning solutions (NH_3).

6.10.3. Periodic checks of water quality

	Unit	Frequency	Max. value
Conductivity	μS cm^{-1}	daily	2 μS cm^{-1}
Total chlorine (including chloramines)	mg/l^{-1}	*	Undetectable (at least <0.02)
Total organic carbon	mg/l^{-1}	monthly	<0.5
Viable bacteria (20°C, 72 hours)	/ml	monthly	(<10^3, <10^4 if stored)
Ammonium	mg/l^{-1}	*	Undetectable (at least <0.05)

* Frequency to be adapted to local condition according to the raw water quality. Additional parameters may also be required for the same reason (fluoride).

6.10.4. Biological suitability test for water

Principle: cultures of *Enterobacter aerogenes* are made in liquid media; one medium is made with the water to be tested, the other with high-quality reference distilled water (e.g. high-quality water for pharmaceutical/injection use, codex guaranteed).

A difference greater than 15 to 20% between the numbers of colonies after 24 hours in the test culture and in the control gives evidence of toxicity (See ISO 9998-1991, Annex B).

This test should be carried out on commissioning of the apparatus and after any subsequent changes. Testing may also be required during the investigation of any quality control problems.

6.11. Culture media and chemicals

6.11.1. Visual control and registration

Upon arrival, each pack of dehydrated or ready-to-use medium or chemical must be inspected and registered. Inspection consists of looking for any abnormality, leakage, clogging, etc. Registration consists of documenting the manufacturer, name of product, date received, quantity, batch number and expiry date. The dates (received, open, expiration) must be reported on the bottle.

6.11.2. Storage

Dehydrated culture media must be stored in accordance with the manufacturer's instructions. In the absence of these, storage should be at room temperature in a dry and dark place.

Ready-to-use media (broths and agars) should be kept in the refrigerator (around +4°C or +6°C) except tightly (screw-) capped tubes and bottles which can be kept in a dark, dry place, at room temperature. Some chemicals such as oxidase reagent must be kept in the dark and refrigerated.

The storage times of prepared media must be validated in the laboratory. Certain constituents such as antibiotics may have short shelf lives and, as the constituents age, recovery and selectivity may be altered.

Autoclaved media should be re-melted only once and should not be kept melted at 45–50°C for more than eight hours before use.

6.11.3. Bacteriological dyes

Dye solutions should not be kept for more than three months. Gentian violet (crystal violet) should be filtered on paper as soon as microscopic examination (Gram's stain) shows particles.

6.11.4. Antibiotics

Solutions should not be kept for more than three months (after being made up). They require the correct diluent and some require alcohol or another organic solvent to give a particle-free solution of the correct concentration. They must be filter-sterilized and must be labelled with batch numbers and expiry dates.

6.11.5. Preparation and dispensing

The medium to be prepared should be selected, according to the written procedure for the intended culture method. The following should be registered in a file or notebook:

(i)   the date,

(ii)   the nature of the medium,

(iii)   quantity of dehydrated product and/or any component,

(iv)   volume of water,

(v)   number and volumes of vials filled (tubes/bottles),

(vi)   autoclaving conditions.

Only controlled items and procedures should be used, namely:

(i)   chemicals and dehydrated media should not be out of date,

(ii)   water should have been tested,

(iii)   containers should be clean, neutral and non-toxic,

(iv)   the autoclaving must be under control and verifiable.

6.11.6. pH measurements

pH measurements may have to be made during preparation or after autoclaving. The final pH prescribed in a procedure must be measured with a calibrated pH meter at room temperature, after cooling the medium. For solid media, the agar is crushed to release the interstitial water for pH measurement. It may be necessary to add some drops of deionized water to provide good contact with the sensor.

The pH cannot be modified once sterilization has taken place. It is only a check of validity and will indicate any pre-autoclaving pH adjustment that may be required to correct any pH drift induced by autoclaving. Adjustments of pH should be made using 1 N HCl or 1 N NaOH. Autoclaving should follow preparation within three or four hours.

Heat-sensitive components of media must be sterilized separately either at a lower temperature or by filtration. Glucose is sterilized at 110°C for 15 minutes, to avoid thermal destruction and Maillard reactions with amino acids and peptones. Vitamins are sterilized by filtration, at room temperature. Phosphate buffers should be sterilized separately from other complex ingredients (such as yeast extracts, etc.) to prevent reaction with Ca^{2+} ions producing insoluble $Ca_3(PO_4)_2$.

To allow steam penetration, the autoclave must not be overloaded and cotton plugs or loose screw-cap closures should be used. Care must be taken to prevent contamination after autoclaving, cotton plugs should be covered with paper or aluminium foil, and the screw-caps should be tightened as soon as possible after autoclaving.

6.11.7. Quality control of prepared culture media

The aim of quality control is to ensure confidence that the target organisms will always be recovered at a known sensitivity. The media should therefore be tested for the characteristic response with reference strains and, if the media is used for counting micro-organisms, additional checks of quantitative recovery and selectivity are required.

All tests should theoretically be performed on each new batch of medium, that is each set of tubes or bottles made from the same batch and autoclaved together. This may increase the workload and it is advisable to:

(i)   batch media preparation for a one week (or two weeks) supply instead of making small quantities every day;

(ii)   use single batches of dehydrated media, thus reducing the need to check the composition repeatedly;

(iii)   obtain and evaluate the results of the quality controls made by the manufacturer for each batch.

The controls made on each batch of dehydrated media cannot guarantee the good quality of each batch of prepared medium. For example, some media require the addition, after autoclaving, of sugars, antibiotics, dyes or indicators and these added characteristics should be checked for each batch of prepared medium.

6.11.8. Sterility testing of media

The test should be done as soon as the batch is put into storage. Ideally, no tubes or bottles from the batch should be used before the result of the test.

It may be necessary to subculture prepared liquid media to verify the absence of growth (turbidity or colonies) in the tubes or bottles, because some can be contaminated without exhibiting growth.

For opaque broths which have been incubated, a loopful should be spread onto non-selective agar to test for colony appearance. The test should involve several tubes or bottles (at least $\log_{10}n$). If only one shows growth, the test should be repeated. Otherwise, any growth should lead to rejection.

The test for sterility should be repeated if any problems arise in the daily use and if the normal storage period is to be extended. Sterility should be tested for each batch of prepared medium.

6.11.9. Other tests: media for isolation or biochemical characterization

A collection of reference strains should be maintained in pure culture in the lab and serve as negative and positive controls for media

Table 6.2

Examples of strains to be used for quality controls

Medium/reagent	Positive control	Negative control
King A medium	*P. aeruginosa* (type A)	*P. fluorescens*
Coagulase reagent	Coagulase positive *Staphylococcus aureus*	*Micrococcus*
Oxidase paper	*P. aeruginosa* or *Aeromonas*	*E. coli*

(and reagents (Table 6.2). These reference strains should be obtained from national culture collections and only sub-cultured on one occasion when many individual sub-cultures in tubes or beads are made and then stored frozen until required for use. This controls changes that could occur in reference cultures and even replacement by a contaminant.

The strain for which the medium is intended should give the typical characteristic described for the original formulation.

6.11.10. Other tests for media intended for micro-organism enumeration

The recovery of the target organism and the inhibitory power against the most frequent interfering organisms (non-target organisms) should be tested on a quantitative basis and used to compare the in-use batch with the new batch (Table 6.3).

Cultures (at least one target and one non-target strain) are made in broth, and they are diluted to produce a usable innoculum (10–100 on small or 30–300 on big Petri dishes). The dilution can be determined by optical density measurements or, alternatively, reference material of a known contamination level can be used. These suspensions are seeded onto the test medium by incorporation, surface spreading or filtration, according to the procedure for use of the medium. Using the same method of inoculation, the suspensions are seeded onto a non-selective medium.

Table 6.3

Examples of organisms to test media

Medium	Target organism	Main non-target organism
TTC-Tergitol Agar	*E. coli*	*P. fluorescens*
Slanetz-Bartley *Enterococcus* Agar	*Enterococcus* sp	*E. coli* *Aerococcus viridans** *Staphylococcus**
Plate count agar Mannitol salt agar	organism *Staphylococcus aureus*	*E. coli*

* Non target organisms to include in the test protocol if the procedure does not involve confirmation.

After incubating according to the procedure, the colonies are counted or the MPN recorded. The recovery of the target organism should be at least 66% of the result on non-selective medium. No non-target colony should appear. Alternatively, the recently available reference materials may be used to provide organism suspensions of known contamination levels (see Chapter 8).

References

Geldreich, E.E. and Clark, H.F., 1965. Distilled water suitability for microbiological applications. J. Milk Food Technol., 28: 351–355.

Chapter 7

Quantitative method and procedure assessment

7.1. Introduction

Nearly all measurements in food and water microbiology are method-dependent and in standardized methods complete instructions for the analysis are given, down to the media composition and its pH. In the selection of methods to be used in a laboratory it is important, therefore, to use published standard methods where they are available. In some countries, the methods are specified in legislation and there is therefore no choice. Standard methods are available from the following sources:

(i)   ISO: International Standards Organization;

(ii)   CEN: Comité Européan de Normalisation (European Committee for Normalization);

(iii)   IDF: International Dairy Federation;

(iv)   AOAC: Association of Official Analytical Chemists;

(v)   Individual standards organizations of each country.

These standard methods have been validated and will be reliable if they are adhered to; if a standard method is not published then one may be available from government laboratories or the major institutions, particularly those who are exploring new methods and improving existing methods.

There have been examples in the past where individual laboratories have changed the standard methods without revalidating them, which has led to an unnoticed deterioration in performance. This chapter describes the types of methods, gives explanations of the factors that describe method performance, and gives details of the procedures that can be used to compare and validate complete methods or their constituent parts.

Accreditation bodies will expect validated methods to be used and if non-standard methods are used they will wish to examine the validation data.

7.1.1. Performance characteristics from a scientific and an operational point of view

The microbiological examination of water and food may be aimed at the detection or quantification of a great variety of organisms including viruses, bacteriophages, bacteria, fungi, yeasts and protozoa. They may be pathogenic or toxinogenic for man or animals, may cause spoilage or otherwise impair the quality of a product or they may indicate whether a product has been adequately processed or, conversely, has become contaminated. The requirements on the performance of a microbiological method may vary considerably depending on the aims of the investigation, the frequency with which it is being repeated, etc. The analyst will, however, often have to be satisfied with a method that does not meet all necessary requirements, simply because a better alternative is not available. When choosing a microbiological method and defining the necessary performance characteristics, these aspects must be taken into account and agreed with the client. Havelaar (1993) has distinguished two types of performance characteristics: the scientific aspects and the operational aspects. The scientific aspects are fully described in Chapter 8. The operational aspects relate to the ability of a method to produce data within an adequate time frame to take appropriate action, to produce data at a reasonable cost and to the complexity of the method in relation to required expertise and equipment. Many microbiological measurements are carried out in monitoring and verification programmes, to evaluate the performance of a process. In such cases there will be high demands on the operational aspects of the analytical method, and it may be necessary to make a compromise with regard to the scientific aspect. On the other hand, if the presence or absence of a pathogenic micro-organism is sought, the demands on scientific quality will generally be high, and the cost and time-to-result of the analysis are of lesser concern. This chapter will summarize the most important aspects of the scientific performance characteristics of microbiological methods, aiming at their application in quality assurance. The emphasis will be on bacteriological culturing methods, because most information is available in this field.

7.1.2. Target organisms

All microbiological methods are designed to detect and/or enumerate particular types of micro-organisms, the target-organisms. All other micro-organisms that may be present in the sample should go undetected and should not interfere with the analytical process, these are the non-target organisms, also described as competitive or background flora.

If a non-target organism is mistakenly identified as a target organism, a false-positive result is obtained; the reverse false-negative result is obtained if a target organism does not give a characteristic or 'typical' reaction in the test. Note that false-positives and false-negatives may be defined for single colonies, but also for the final result of the examination of a sample. The nature and concentration of non-target organisms and target organisms may vary considerably between samples, certainly from different locations, but also in time at a certain location. This implies that a method that has been evaluated for a particular type of sample does not necessarily have universal applicability. To overcome this problem, methods have been laid down in (inter)national standards or legal requirements, but even then the laboratory remains responsible for evaluating the performance of the method for the type of samples under investigation, and for seeking alternatives when necessary and possible. The possible temporal variation of the performance of a method in relation to variable characteristics of the microflora should be evaluated as a part of the quality assurance programme.

7.1.3. Method-defined parameters

Ideally, the target organisms should be defined in taxonomical terms, i.e., a particular (group of) species or genus. A common practice in water and food microbiology is, however, to define the target organisms by the method that is being used for analysis. This practice may make the evaluation of the performance of a method, and the comparison of different methods difficult. This may make it difficult to replace an outdated method by a more recent and better one, particularly if the original method has been standardized or been included in the law.

The results of a microbiological test are always method-defined, i.e. when examining the same sample by different methods, different results may be found. This is particularly so for detection of target organisms that are not unequivocally defined by the method, but also for taxonomically defined target organisms. The viability of a micro-organism may affect detection; damage to micro-organisms can vary in degree and type. Different methods will recover different proportions of the population, for example the comparison of microscopic counts with plate counts on highly selective media. There is no objective criterion to define which method gives the 'better' result, which will also depend on the aims of the analysis. Because results are method-defined, great care should be taken in standardization of methods, but this until now has not been done to a sufficient extent.

7.2. Principles of microbiological measurements of water and food

The following paragraphs give a brief overview of the most common methods used for detection of micro-organisms in water and food. Recent reviews, where more details can be found, are Havelaar (1995) for traditional methods, Karwoski (1994) for automated methods, Morgan *et al.* (1992) for immunological methods and Olsen *et al.* (1995) for molecular methods. Reference is also made to the Micro Val project, which aims to define validation criteria for test-kits in food microbiology, and to evaluate commercially available products.

7.2.1. Enrichment in liquid media

In liquid enrichment methods, a test portion is inoculated into a growth medium that has been formulated to stimulate growth of the target organisms and to suppress growth of all other organisms (background flora). The selective nature of the enrichment medium is enhanced by choosing an appropriate incubation temperature and time. If the target organism is present in the test portion, this will usually result in a positive signal, irrespective of the original number. In its simplest form, a liquid enrichment method therefore gives a presence/absence type of information. In order to obtain (semi-)quantitative information, a series of different volumes (e.g. 100, 10, 1 and 0.1 ml) may be examined to produce an end-point type of result. If a series of different volumes is examined in replicate, e.g. three- or five-fold, it is possible to use a statistical method known as the 'most probable number' (MPN) technique to estimate the original concentration of the target organism. The precision of this estimate is low (e.g. the 95% confidence interval of a five-fold MPN estimate is roughly between one-third and three times the analytical result), and some people hesitate to call MPN methods more than semi-quantitative. Imprecision in MPN, however, is not inevitable but due to the small number of parallels normally employed. If the number of parallel test portions is increased to 100 or more, the MPN technique will surpass the conventional plating technique in precision. The simplest technical solution to increasing the number of parallels has been achieved in the hydrophobic grid membrane filter technique where 1600 growth cells are inoculated in a single filtering operation. The filter is subsequently incubated on a solid medium. Another, similar single-dilution MPN procedure is based on the Anderson air sampler (400 parallel growth cells). A somewhat less convenient way to increase precision is to use microtitre plates of 96 wells together with multitip pipettes (Hernandez *et al.*, 1991). This technique permits the use of different dilutions, thereby increasing the counting range. Precision of single-dilution or multi-dilution MPN estimates is inversely related to the square root of the number of parallel tubes or growth cells.

7.2.2.   Colony counts

In the colony count method, a test portion is inoculated onto the surface of a growth medium that has been solidified by addition of agar–agar (spread-plate method). Each individual cell of the target organism will multiply into a colony that is visible to the naked eye. If several cells of the target organism are physically connected (e.g. by adsorption to a particle of suspended matter) this will also result in one colony. The results of the plate count technique are therefore expressed as the (number) concentration of colony-forming particles (cfp) per unit volume. Each cfp represents one or more cells of the target organism in the original sample. Variations are the pour-plate method where the test portion is mixed with the liquefied agar medium, poured into Petri dishes and incubated after solidification, and the membrane filtration method where the test portion is filtered through a membrane filter (usually of 0.45 μm pore size) and the filter is placed on the growth medium.

It is sometimes not sufficient simply to inoculate a test portion in or on a selective growth medium to obtain accurate results. Cells of target organisms may be damaged to some degree by physical or chemical stress and it may be necessary to revive them before placing them in the selective environment of the growth medium. This procedure, called resuscitation, may be an integral part of the test method and usually involves incubation in a less selective medium and/or at a less restrictive temperature.

The selective growth medium is usually supplemented with some kind of specific detection system to differentiate growth of target organisms from that of background organisms. The detection system can be based on fermentation of specific sugars, enzymatic degradation of specific substrates, mobility, reduction of hydrogen acceptors, etc. and will usually result in recognizable colour changes, gas production, etc. Routine methods are designed to produce results with an acceptable degree of selectivity at this point. For more critical examinations (e.g. detection of specific pathogens), it may be necessary to examine further the 'presumed' positive results by one or a series of confirmatory tests or even to proceed to identification at the genus, species or type-level. The objectives of the analysis determine the necessary level of confirmation and identification.

7.2.3. Microscopical methods

Direct enumeration of micro-organisms by microscopical methods has limited applications in water and food microbiology because the detection limit is relatively high and because the microscopical image provides only a marginal clue to the identity of bacteria. This latter fact can theoretically be overcome by immunofluorescence methods, but only a few selective antibody preparations are readily available. Furthermore, microscopical methods do not differentiate between living and dead cells, which makes interpretation of analytical results in terms of health risks impossible. Several methods have been developed to assess the viability of single cells by microscopical methods, such as exclusion of certain dyes (indicating integrity of the cell wall), reduction of tetrazolium salts (indicating active respiratory metabolism) and cell elongation in the presence of nalidixic acid (indicating active biosynthesis). These methods are laborious and require the expertise of a research laboratory. It has been demonstrated that in a bacterial population under stress, the detectability by culture methods is lost more readily than the viability using microscopical methods and claims have been made that bacteria in this viable, non-culturable stage are infectious to man and experimental animals. However, these claims have not been supported to any great extent by published material and solid contradictory evidence has been published. The infectivity of viable, non-culturable bacteria presently remains the subject of considerable debate.

Some examples of direct microscopical methods that do have application in food microbiology are the direct epifluorescence test (DEFT) to quantify bacteria in milk or haemodialysis water, and the Howard mould count to detect fungi in food products. Such applications aim at producing total counts, i.e. no attempt is made to differentiate the bacteria or moulds seen under the microscope.

There is growing interest in the occurrence of the protozoa *Giardia* and *Cryptosporidium* in (drinking) water, which is detected by immunofluorescence microscopy. Ideally, only infectious oocysts of the human pathogenic species *G. lamblia* and *C. parvum* should be detected and differentiated from non-infectious oocysts or those from other species.

7.2.4. Impedimetric methods

The multiplication of micro-organisms results in changes in the chemical composition of the growth medium. These changes can be measured as a shift in the electrical resistance (impedance) of the medium, that can be recorded by special, commercially available apparatus. The number of micro-organisms present in the inoculum can be estimated from the rate of change of the impedance, particularly from the time interval between inoculation and the detection of a significant change of the impedance. The success of impedimetric methods depends entirely on the selective properties of the growth medium. The first uses of impedimetry were to replace general parameters, such as total plate counts, sterility testing, yeasts and moulds, etc. Selective media for detection of groups such as the coliform bacteria, pseudomonads and enterococci have also been designed. Methods for specific pathogens are being developed, a method for *salmonella* spp. in foods having been recognized by AOAC.

7.2.5. Immunological methods

Immunological methods have found widespread use in food microbiology; a recent overview of methods and applications can be found in Morgan *et al.* (1992). Applications include detection of pathogens such as *salmonella* spp., *Listeria monocytogenes* and enterotoxigenic *Clostridium perfringens* (usually after pre-enrichment in a suitable culture medium), and detection of bacterial and myco-toxins. A diversity of methods is used, such as enzyme immunoassays, immunomagnetic separation methods, agglutination methods, etc. Products may either be available as specific reagents, or as easy-to-use complete test kits, dip-sticks, etc. The quality assurance and validation of such methods is rapidly developing.

7.2.6. DNA methods

The potential of molecular methods for detection of bacteria in water samples is currently being explored in many research laboratories. Direct hybridization assays have limited applications because of the high detection limit. The polymerase chain reaction (PCR), in which single DNA fragments are serially amplified by *in vitro* enzyme reaction, holds great promise but needs further development before large-scale applications can be made. PCR is basically a presence/ absence technique but it could be used in an MPN-type of assay to produce semi-quantitative results. A major area of study lies in the differentiation between living and dead cells. A simple, practical solution is to precede the PCR step by a short enrichment step allowing culturable bacteria to multiply. This may also partly solve the problem of interference of the PCR reaction by constituents of the sample matrix.

7.2.7. ATP measurements

Adenosine 5-triphosphate (ATP) is present in all living cells, but is rapidly broken down once a cell dies. Bacteria contain about one femtogram ($= 10^{-15}$ g) of ATP, yeast cells 10–100 times more. ATP can be detected by the lucifer–inluciferase system, using highly sensitive photomultipliers to detect the light emitted. ATP-measurements can be used directly to quantify bacterial activity in water or biofilms; they are also used to measure the biofilm growth potential of (materials in contact with) drinking water (Van der Kooij and Veenendaal, 1992; 1993). For analysis of food products, the sample needs to be pre-treated to remove somatic non-microbial ATP. Such pre-treatment may be complicated, and may affect the detection limit of the method. Some successful applications are total plate counts in meat samples, quality control of fruit juices and prediction of shelf life of food products.

7.2.8. Turbidimetric methods

Growth of micro-organisms leads to an increase in the turbidity of the medium which can be detected by optical instruments. It is possible to measure changes in absorbance, or in light-scatter. Several instruments are commercially available. Applications have as yet been limited to sterility testing (e.g. of heat-treated milk) or replacement of total plate counts. It is also possible to use turbidimetry in bacterial identification systems.

7.3. Standardization and validation

The long-term aim of standardization and collaborative testing is to harmonize methods across the world. So far, this end has not been completely achieved. Methods for the same purpose (target) differ in different countries and even between laboratories of the same country. It is entirely possible that the variation is justified by experience; differences in the microbial populations in different samples and in different geographical areas may require different methods.

It is somewhat disquieting that the use of different methods seems to follow cultural and language borders rather than geological and ecological boundaries. Our knowledge in this area is inadequate. It would be important to know how much divergence is really necessary.

(a) Present situation

In some cases, standards (e.g. ISO) allow variation as to the choice of nutrient media, inoculation technique (plating vs. surface spreading or spiral plating, etc.), incubation temperature (35 vs. 37°C, 44 vs. 44.5°C) or analytical principles (colony count vs. multiple tube). Laboratories may need to validate their choice on experiments with their own range of samples. Moreover, laboratories may have questions related to the influence of their equipment on the results.

(b) Purpose and scope

Normally, the daily analytical routine provides no data for method evaluation. Chapter 7 aims to present some procedures for method assessment so that confidence can be gained in good methods and so that methods found to have poor performance can be replaced.

There are two broad problem areas. One is the choice of an optimal analytical procedure when there are alternatives. The other is quality control of the procedure of choice. The use of different types of replication for QC purposes in daily routine examinations is presented mainly in Chapter 8.

Validation of a method is a formidable collaborative research task and cannot be tackled in this guide. Single laboratories occasionally face the problem of implementing a new method or a procedure. This chapter provides some designs for such purposes.

The purpose of Chapter 7 is to give guidance on experimental designs to microbiologists and an overview of microbiological quantitative methods to statisticians so that method evaluations can be planned. It deals only with the quantitative aspects of testing.

The effects of the bacterial population, nutrient medium and technician are almost inseparable. It is seldom possible to assess the performance of a method separately from the analyst performance. The matters dealt with in Chapter 7 are closely related to the contents of Chapter 8, which focuses on quality assurance.

(c) Methods

The definition of method adopted by the chemists is so general that it suits microbiology as well. The final draft of the document CEN TC 230/WG 1/TG 4 'Guide to analytical quality control for water analysis' contains the definition: "The analytical method is the set of written instructions followed by the analyst". Two analytical procedures that differ in any detail could therefore be considered different methods. Differences in details of sample dilution are, however, not normally considered sufficient reasons to call methods different in microbiology. Instead, differences in the formula of the nutrient medium, the detection principle, and time/temperature of incubation make the methods different.

(d) Good methods

A good method is precise, robust and specific to the target. It gives the correct result on the average (trueness). Robustness implies stability. A robust method is insensitive to the small variations in analytical performance and laboratory environment that unavoidably remain, even in well-controlled laboratories.

(e) Generation of method assessment data

Methods are most effectively assessed with specially designed experiments. Data cumulated in the course of routine work are usually too imprecise to serve the purpose very well. Besides, routine analyses usually lack a plan of replications that is a prerequisite of method comparisons.

There is seldom a reason to compare microbiological methods using low analyte densities. Even presence/absence (PA) methods that are designed to operate at low microbial densities can, at least initially, be efficiently evaluated at such densities where absence of the target organism from the sample portions studied is highly unlikely (see Section 7.4).

Whenever general statements are required, the conclusions must be based on an assortment of natural samples.

(f) Method performance tests

In method performance tests, a group of selected laboratories examines a number of relevant materials using an exactly specified method. Each laboratory receives 'identical' blind duplicate sub-samples of each material. This procedure has been successfully applied in chemistry to obtain precision estimates (repeatability and reproducibility) of various methods.

In microbiology, similar performance tests have not been widely applied and have not always given equally promising results. The repeatability and reproducibility estimates obtained in collaborative tests are frequently very high. Reasons may be contagious distributions of microbes in sample materials and lack of analyst performance assurance. Considerable systematic components may remain. Carefully controlled certification and intercalibration studies have, however, produced repeatability and reproducibility values close to theoretical optimum values. With more experience, collaborative method performance studies will most probably become as useful in microbiology as they are in chemistry, although methodology will have to take account of the differing levels of detection expected in microbiology (one organism) compared with chemistry (usually thousands of molecules) (Tillett and Lightfoot, 1995).

(g) Modern and traditional methods

The aim of microbiological analyses is usually to furnish a quantitative measure of the quality of food or the environment, in most cases by studying samples. This aim can be approached in different ways.

Some so-called rapid, modern, or 'routine' methods assess product quality by dynamic measurements (lag time, time of detecting a change in colour, turbidity, electrical conductivity, growth rate, maximum turbidity, etc.). The traditional methods aim at estimating the 'standing crop' of viable target organisms by their colony-forming ability, given adequate time. It is not microbiologically necessary for the signal detection time of the modern methods and the viable count of the traditional methods to be exactly related.

The modern methods have the advantages of speed, economy, and high capacity due to automation and miniaturization. In large laboratories there is therefore considerable interest in adopting these techniques. Smaller laboratories prefer to forgo the investment and stick to the classical methods. Reconciliation of the different analytical concepts is largely unsolved at present. The Micro Val project (see also Section 7.2) is working on a solution for validation of the new methods and the results should be available soon. Also, IDF working party E29 has worked on the problem for several years and a provisional standard is expected to be published soon. It approaches the comparison of reference and 'routine' methods, principally with the help of visual inspection of correlation and regression data.

7.4. Mathematical and technical characteristics of microbiological methods

(a) General

Strictly speaking, all microbiological culture methods are selective. They cannot provide suitable growth conditions for the whole spectrum of micro-organisms present in a natural sample. The number detected is always only a part of the whole (usually much less than 10%). Nevertheless, methods where selectivity is unintentional are usually called non-selective. They aim at maximizing the number detected.

Non-selective methods are mathematically straightforward. The number of colonies is simply assumed to be an accurate representation of the number of those vital (able to be cultivated) particles in the test portion that are capable of development under the conditions provided.

Intentionally selective methods aim to quantify a defined sub-set (target population). The practical solutions usually involve both suppression of the often dominant non-target population and detection of the target colonies by characteristic growth reactions. The latter aspect creates the situation where human analysts must be able to differentiate between typical and non-typical colonies when reading the results. Harmonization between individuals, and especially between laboratories, needs further development.

(b) Mathematical considerations

The classical quantitative estimation methods in microbiology make use of two types of detection principles. One is to detect and count individual particles; the other is based on detection of mere presence of micro-organisms in a test portion.

Assume that g is the number of active growth entities (viable particles, germs) in the test portion. When they develop in or on a solid matrix, each one potentially gives rise to a colony. The number of colonies counted is often called the number of colony-forming particles (formerly colony-forming units, CFU, cfu), meaning that it is assumed to be equal to g. In reality there are many reasons why the colony count observed often is not equal to g. Abundant growth of background colonies may inhibit or mask the development of the target colonies, simple crowding may lower the count, or accidental errors in growth conditions (humidity, temperature, medium) may change the count radically.

In methods based on presence/absence detection, the sample containing the g growth particles may be subdivided into several smaller test portions (equal or unequal) to obtain a numerical estimate. The estimate of g is based on probability calculations. By convention, the point of maximum probability (mode of the probability density curve) is taken as the estimate of g. Hence the term MPN (most probable number) is used for these estimates. Due to skewness of the probability distributions, the MPN estimate is not equal to the arithmetic mean.

For the time being, the modern methods are not independent and have no mathematical theory comparable to the Poisson distribution that is the basis of both the MPN and colony count methods. Modern methods are calibrated against the colony count estimate. Ultimately, the goal should be to free the modern methods from the necessity of validating them with the colony count and to relate the dynamic automated measurement directly to product or sample microbiological quality.

(c) Uncertainty in measurement

Typically, errors are classified as systematic and random. However, the classification is flexible. Every person develops a style of counting colonies that systematically gives low or high results compared to another person. When several laboratories join a collaborative performance test, the systematic counting 'errors' appear as an unaccountable random error when the data are considered as a whole.

The two classical microbiological detection principles function only at low particle densities, at most in the hundreds per test portion. At such densities, considerable random variation appears between particle counts, even between accurately and correctly measured parallel test portions. Such uncertainty cannot be lessened by refinement of working techniques, but only by increasing the number of particles counted. This problem exists also in the counting of radioactive particles and has recently caught the attention of chemists, when picomolar concentrations in microlitre volumes are being measured. The numbers of atoms or molecules per test portion are then in the hundreds of thousands. The third or fourth significant figure is uncertain.

In microbiology, the same type of variation affects the first or second significant figure.

7.4.1. Accuracy of microbiological measurements

According to ISO 6107-8 (ISO, 1993), the accuracy of a single test result is defined as the degree of similarity between the measurement and the true value of the measured quantity. The accuracy is inversely related to the overall uncertainty of measurement, that is the difference between the measurement and the true value. The overall uncertainty of a single measurement is thought to consist of two types of components, systematic (i.e. bias) and random (imprecision):

UNCERTAINTY = BIAS + IMPRECISION

This is a simplification of the situation. In microbiology there are usually also contributions to uncertainty that do not fit this classification — unexpected fluctuations that cannot be modelled mathematically. The inverse of bias is the trueness of the method. It is a characteristic of the method used for the analysis and also depends on the nature of the sample. Trueness is not related to errors in applying the method in individual laboratories. These are included in the random error of the measurement, introduced by variations between laboratories or within a single laboratory. The inverse of random error is the precision. Hence, the same relation as above is more commonly expressed as:

ACCURACY = TRUENESS + PRECISION

Inaccuracy is caused by many factors. Its nature in microbiology is discussed below.

7.4.2. Elements of trueness

(a) Recovery

The physiological state of bacteria in water lies on a continuous scale between fully vital and completely dead. In the international literature, the following stages are generally distinguished:

(i)   vital, i.e. capable of growing and/or producing colonies on conventional selective culture media;

(ii)   sub-lethally injured, i.e. capable of growing in/on non-selective media but not directly on selective media;

(iii)   viable, non-culturable, i.e. morphologically intact with demonstrable metabolic activities such as respiration, synthesis of cell material or micro-colony formation, but not culturable on conventional (non-selective) media;

(iv)   dead, i.e. morphologically intact but without any demonstrable metabolic activity.

Different methods may recover different proportions of the total population, thus introducing a method-specific bias that may be very apparent in a highly injured population, but barely noticeable in a population of vital cells.

To test recovery, a standardized stress is applied to a sample spiked with the target organism. The ratio of colony numbers under selective and non-selective conditions provides an estimate of recovery. Many points should be considered in order to produce meaningful data. These relate to the use of pure cultures or naturally contaminated samples, to the type of injury applied to the test samples, to the preparation of 'homogeneous' (random distribution) and stable, yet representative test samples, etc. It is clear that there is no single figure that can characterize the recovery of a certain method, not even for a particular type of sample. Nevertheless, it is important to study the recovery when validating a method for a particular use, and to constantly monitor the recovery of a standard sample when applying a method.

Bacterial colonies on plates or membrane filters may influence each other and in general there will be a tendency to lower test results if more colonies are found on the same plate. To which extent these so-called crowding effects will influence the final test result will be different for each method, depending primarily on colony size. An elaborate discussion of this problem was published by Niemelä (1965). Furthermore, the linearity of the method can be affected by the ability of the analyst to distinguish 'typical' from 'atypical' colonies. Experience has shown that misinterpretation and differences between several analysts are more likely to occur at higher colony densities, particularly with methods that require subjective interpretation of colours or dimensions of colonies. A laboratory should be aware of the characteristics of their methods and pay proper attention to eventual non-linearities. This will result in a method-specific upper counting limit per plate, rather than the often applied rule of thumb of 300 colonies on a 9 cm Petri dish or 80 colonies on a 50 mm membrane filter. It may even lead to laboratory-specific counting ranges, related to the nature of the samples under investigation and/or the experience of the analysts.

Despite the introduction of automatic devices for colony counting, much of the final interpretation of microbiological tests relies on the ability of the human eye and mind to distinguish minor variations in colour, colony morphology, colony overlap etc. This introduces the risk of subjectivity in the interpretation of results and all possible efforts should be made to standardize the readings of different analysts as much as possible. One simple tool is to select a series of plates and let all personnel involved in the interpretation of test results count these plates twice by blind replication. This will allow the evaluation of the consistency of judgement by the same person and by different persons. Any difference in interpretation should be thoroughly discussed and, where necessary, further confirmatory or identification tests should be done to find the right interpretation. This comparison of different analysts should be repeated periodically and should also be a formal step in the approval of new personnel.

(b) Trueness

In many cases, including the bacteriological examination of water and food, the true concentration of a component is not known and can only be used as a hypothetical aim. The systematic error can therefore not be established in an absolute manner. It can theoretically be approached by studying the same sample repeatedly, using different methods. Because microbiological methods are destructive in nature (i.e. the sample is lost during analysis) this requires well-mixed test materials with randomly distributed bacteria and/or a large group of laboratories. One could use different selective media for estimating the 'true value' of a target organism in a naturally contaminated sample or selective and non-selective media for pure cultures. However, this approach may often not be helpful because different media have different systematic errors. In this respect, the pragmatic approach by ISO/REMCO (document N263, November 1992) is more helpful. Instead of the true value, an accepted reference value is used to define trueness.

For microbiological methods involving a cultural step, several aspects of trueness can be recognized. These are related to the differences in recovery (quantitative errors) and the differential characteristics (qualitative errors) of a test method in relation to both target and non-target organisms. They are summarized in Table 7.1. For other methods the same principles can be applied, but have not yet been worked out in detail.

Table 7.1

Trueness relationships for target and non-target organisms

	Growth characteristics	Differential characteristics
Target organisms	Recovery	Sensitivity
Non-target organisms	Inhibitory power	Specificity

7.4.3. Elements of precision

General principles for the characterization of the precision of test methods have been described in ISO 5725 (1994). Two measures of precision are described in this standard, repeatability and reproducibility. The formal definitions are as follows:

"Repeatability: the closeness of agreement between mutually independent test results obtained with the same method on identical test material in the same laboratory by the same operator using the same equipment within short intervals of time."

"Reproducibility: the closeness of agreement between test results obtained with the same method on identical test material in different laboratories with different operators using different equipment."

(Note that in microbiology truly identical test material cannot be supplied. A random distribution of the micro-organisms is the best possible objective, and the degree to which this is reached is critical for establishing r (repeatability value) and R (reproducibility value). Any over-dispersion in the test materials will add to the variability of the test results.)

There are various factors that contribute to random errors (using 'error' in the statistical sense) of microbiological counts, which thus affect repeatability and reproducibility. These are considered in more detail below.

(a) Repeatability

Quantitative microbiological analysis is applied to samples of water and food that should be representative of the object under investigation and that should be well conserved to maintain all relevant properties constant until the time of analysis. This subject is covered in other chapters. The analytical process *per se* usually starts with homogenization (i.e. thorough mixing) of the sample, taking one or more sub-samples, diluting or concentrating these sub-samples when necessary and plating one or more replicates from each dilution or concentration step. Under repeatability conditions, two major sources of error can then be recognized — the variation between replicates from one dilution step and the variation between different sub-samples (including dilution or concentration errors). These errors include both a random component due to the intrinsic variation of the number of microbes in small volumes of test samples and a component related to the precision of the volumetric glassware used, the attitude of the analyst, etc. In general, data from microbiological counts are not normally distributed. At the level of replicates from a homogeneous test solution, the numbers of colony forming entities may be expected to follow a Poisson distribution (Hildebrandt *et al.*, 1986). Test results from different sub-samples may in the ideal case also follow a Poisson distribution, but will often show more variation than can be expected on this basis alone because additional sources of error must be taken into account, such as dilution errors and the effects of particular or colloidal matter on the distribution of microbes. Experience has shown that, for pure culture suspensions, the data can usually be characterized by log-normal distributions, provided the average count is not too low. If data are derived from natural samples, other distributions such as the negative binomial may be more appropriate (Haas and Heller, 1990).

(b) Within-laboratory reproducibility

Within-laboratory reproducibility is an important target of internal quality assurance programmes. It guarantees that a laboratory is able to produce consistent results through time. A useful tool for the assessment of this factor is to analyse standard samples with each series of analyses. This requires the availability of easily accessible, homogeneous and stable samples. Ideally, this concept leads to the availability of standard reference materials. They are currently being developed in the form of gelatin capsules filled with spray-dried milk containing a well-characterized test strain as described in Mooijman *et al.* (1992).

There are other, less sophisticated possibilities to produce standardized cell suspensions for QA, that can be done by individual (or small groups of) laboratories. These may be quite useful for internal QA programmes or for inter-laboratory tests in limited geographical regions. One possibility is to prepare a standardized suspension of a micro-organism in skimmed milk, which is quickly frozen in ethanol dry ice and stored at $-70°C$. After thawing, 1 ml aliquots can be directly withdrawn and analysed. The third possibility is to culture the organisms in minimal medium and to store this culture in the refrigerator. Normally, the contamination level will remain constant during several months and dilutions to the proper level can be made when desired (Schijven *et al.*, 1994).

(c) Between-laboratory reproducibility

Between-laboratory reproducibility is important to enable comparisons of results from different origins. The factors that determine the variability of results between microbiological laboratories are largely unknown. It has been the experience in a great number of collaborative studies with European water and food laboratories that significant differences in results with carefully prepared standard samples could only rarely be attributed to recognized technical differences, despite extensive questionnaires on details of the procedures as followed in the individual laboratories. Partly, this may reflect the need for introduction or improvement of internal quality assurance programmes, but it also reflects the lack of knowledge of which factors are really critical. More recent experience in certification studies, involving a selected group of laboratories that had worked together for several years, did produce encouraging results, however. Identified deviations from the specifications in the analytical protocols did lead to deviating results, but results obtained under specified conditions showed a high degree of conformity, and no outlying data. Continuous attention to inter-laboratory comparisons is therefore an essential step in the process of improving and standardizing microbiological test results. Two tools are most useful for this purpose: proficiency testing schemes and the use of certified reference materials (CRMs).

7.4.4. Robustness

The robustness of a method relates to the stability of the results towards changes in the testing environment (including physical, chemical and personal factors). Lack of robustness increases variation and decreases reliability of the analyses. Empirical data on robustness do not accumulate during the daily routine analyses. Special tests are needed.

The statistical testing of factors affecting robustness (ruggedness) of chemical analyses has been formulated by Youden and Steiner (1975). It involves selecting seven potentially effective factors and observing their effects at two levels within the normally expected limits. This principle would be admirably suited for microbiology as well but seems never to have been applied. Its prospective principal use would be in the prenormative testing of new methods, and of old ones when existing standards are being reviewed.

In the microbiological context there are some important factors of robustness that are difficult to control and quantify. The first is that selective methods vary greatly in their sensitivity to variations in the non-target population of the sample. Secondly, methods vary as to how easy or difficult it is to read the target population under the influence of the non-target population. Besides, the two factors are interconnected. The same non-target colonies may interfere with the results of one analyst but not of another. The existing quality control methods and method performance tests do not cope with these problems. Chapter 8 deals to some extent with ways of detecting the influence of these factors in numerical data.

7.5. Selective methods

(a) Introduction

Selective methods are designed for detecting and enumerating a defined sub-set (target organisms) in a microbial population consisting mainly of non-target organisms. The often harsh conditions needed to suppress the non-target group may reduce the recovery of the target population as well. Besides, non-target organisms will not always be totally eliminated and the characteristic appearance of the target colonies may not be unequivocally distinct. Few selective methods can be trusted to function so well that verification of the primary colonies is unnecessary in all sample types.

(b) Inhibition of background flora

High densities of non-target organisms on membranes or in enrichment cultures may strongly affect the results of a microbiological analysis. This may be related to crowding effects, competition for nutrients, production of inhibitory substances, neutralization of pH changes by target organisms, etc. As a general and simple tool, it is advisable to note the background flora on plates or membranes so that aberrant data can eventually be retraced. In general, membranes or plates should never show confluent growth of bacterial colonies and even when isolated colonies are seen, care should be taken to ensure that there is no mutual interference.

The possible negative effects of background flora can be quantitatively assessed by spiking samples with standardized suspensions of target organisms or by examining the linearity of the method over a range of dilution steps. Designing and interpreting spiking experiments may be relatively straightforward if the sample is not expected to yield colonies of the target organisms (e.g. drinking water), but may be quite complicated if the target organism does occur (e.g. in sewage). For this latter case, experimental and statistical protocols need to be established. The examination of linearity will be discussed in more detail in a later paragraph.

(c) Inclusiveness and exclusiveness

The inclusiveness of a method defines how far all target organisms that are able to develop colonies under the test conditions do indeed give characteristic reactions, whereas the exclusiveness describes the extent to which colonies of non-target organisms do not exhibit these properties. The use of inclusiveness (I) and exclusiveness (E) to evaluate the differential characteristics of a method is a more direct measure of its performance than the commonly applied verification rates of 'typical' and 'atypical' colonies. The calculation of I and E can be based on the verification rates as shown below.

7.5.1. *Numerical characterization of selective methods*

The characteristics of selective methods can be calculated from the results of experiments, where the total number of colonies (C) is first subjectively divided into two mutually exclusive groups, 'typical' i.e. presumptive target colonies (C_t) and 'atypical', i.e. presumptive non-target colonies (C_n). Accordingly, $C_t + C_n = C$. If possible, all colonies should be sub-cultured to verify their identity as target or non-target organisms. If not possible, a random sub-set of C should be isolated.

Following the notation of Havelaar *et al.* (1993), let the verified number of target organisms be T_t and T_n in C_t and C_n, respectively. The numbers of non-target organisms in the two groups are $N_t = C_t - T_t$ and $N_n = C_n - T_n$. Let $T_t + T_n = T$ and $N_t + N_n = N$. For a summary, see Table 7.2.

A large number of descriptive numerical characteristics can be obtained from the table. Some examples are:

(i)   Index of selectivity, according to ASTM, equals C_t/C;

(ii)   Inclusiveness (Havelaar *et al.*, 1993) $I = T_t/T$;

(iii)   Exclusiveness (Havelaar *et al.*, 1993) $E = N_n/N$;

(iv)   Verification rate $p = T_t/C_t$;

(v)   False-positives $= N_t$ (absolute), N_t/C_t (relative);

(vi)   False-negatives $= T_n$ (absolute), T_n/C_n (relative).

Table 7.2.

	Typical colonies	Non-typical colonies	All colonies
Target organisms	T_t	T_n	T
Non-target organisms	N_t	N_n	N
All organisms	C_t	C_n	C

If all colonies cannot be sub-cultured, a random pick of colonies should be made. The performance characteristics are calculated from the sub-sets isolated and verified.

To calculate I and E from the results of sub-culturing both typical and non-typical colonies, assume that the fraction confirmed as target organisms was f_t ($= T_t / C_t$) and f_n ($= T_n / C_n$), respectively. I and E can then be calculated as follows:

$$I = C_t f_t / (C_t f_t + C_n f_n) \quad \text{and} \quad E = C_n (1 - f_n) / (C_t (1 - f_t) + C_n (1 - f_n))$$

In many test protocols, it is recommended to pick a number of typical colonies and to verify their identity. The count result of typical colonies is then multiplied by the fraction of colonies which confirms as the target organism to obtain the final test result. There are several reasons why this practice should be avoided. Picking and examining colonies adds substantially to the amount of work and thus the cost per analysis. Also, the examination should not be restricted to typical colonies but should also include non-typical colonies, which should be counted separately. Because of the amount of work involved, the number of colonies examined is usually restricted to three or five or, at maximum, 10 per sample. This limitation severely influences the precision of the final test result; it can be shown that the precision of the test increases with the absolute number of colonies that finally confirms as the target organism (i.e. number picked × verification rate). Therefore, it makes little sense to relate the number of colonies to be verified to the original number of colonies on the plate or membrane (e.g. 10% or approximately the square root of the number of colonies).) Rather, a fixed number of colonies should be picked, depending on the verification rate to be expected and the desired precision. If low verification rates are frequently encountered in practice, alternative methods should be selected or developed, rather than spending energy and money on the examination of hundreds of colonies.

Verified recovery among presumptive target colonies can be defined as $x_t = kC_t / n$, where C_t = number of typical (primary) colonies, n = number picked for verification, and k = number verified.

This is the estimate typically given as confirmed or verified count although it ignores the possible target organisms among the atypical colonies. Its precision depends to a large extent on the precision of the estimate of the verification rate and therefore on the numbers isolated and verified.

Calculation of exclusiveness does not appear to be often attempted with colony count methods, whereas verification of negative tubes is frequently practised in the comparison of multiple tube methods.

Both verification rate and false-negative rate depend on environmental and personal factors to such an extent that they are not simple and straightforward method performance parameters. For that very reason, they form a significant element in a method's robustness evaluation.

Apart from tests of recovery, there are no clear target values for the method performance characteristics. The control, therefore, consists of recording observations on control charts and noting any unusual deviations. Due to many uncontrollable factors (operator bias, composition and state of the sample bacterial population, medium performance), the factors are difficult to assess separately.

7.6. Comparison and validation of presence/absence tests

Presence/absence (PA) tests have a significant place in the detection of pathogens and indicators. New methods claiming to have an improved recovery or other superior properties regularly appear in the literature. They need to be compared with one another and with established methods.

Technically, the simplest design is to examine a large and representative selection of samples in parallel with the different methods. This is particularly suitable for water. A sample of one litre volume or more can be mixed well enough to give reasonable assurance of random distribution of particles in the parallel sub-samples. A quantitative membrane filter test can be included as a reference, because it allows the study of 100 ml water samples. A good example of such a study is reported as phase 2 of a study reported by the Public Health Laboratory Service (Lightfoot et al., 1995).

For statistical efficiency the best density of organisms for detecting differences between PA tests is 0.69 per volume examined. This is calculated from Poisson probability and will lead to half positive and half negative results on average assuming random distribution of organisms and correct test results. The volume being examined may be 100 ml with membrane filtration or 2 ml with microtitre methods.

In theory, this comparison of PA tests can be achieved with a very few original samples of high-level contamination. These are diluted down to give the approximate desired density and then examined with many replicates so that there are sufficient observations to draw conclusions. The estimated number of observations required can be obtained from standard statistical formulae, having stipulated the size of difference in performance which is to be detected and the acceptable probability levels attached to failing to detect a difference which does not exist.

This approach with a large number of replicates from a few original batch samples may be used as a preliminary stage of comparison and it may quickly demonstrate the inadequacy of a poor PA method.

If the PA tests appear comparable, but there are doubts as to whether the conclusions can be extrapolated to all the samples that a laboratory is likely to test, then another stage of comparison should be considered. This would use a large number of natural samples where PA testing is the issue of importance (e.g. drinking water) which will be typical of the area served by the laboratory. It requires a large number of low-contamination samples to accumulate evidence. The study of coliform organisms referred to above (Lightfoot *et al.*, 1995) found this additional stage very worthwhile. The geographical variation in conclusions became apparent only when these 'real-life' samples were studied.

As pointed out in the study there are two conditions for a successful trial:

(i)   the number of samples examined must be large (over a thousand) even in the most favourable cases;

(ii)   only samples with low but non-zero bacterial content (about five viable particles per 100 ml) are suitable.

If these conditions are not met, the results may remain inconclusive and differences between methods may remain undetected.

Method comparisons in the above design are based on statistical evaluation of frequencies of positive and negative samples. The design is applicable not only to water samples but solid foods as well, because non-parametric statistical techniques are used for evaluation and there is no compelling need to assume a Poisson distribution. Demonstrating that different methods may not work equally well in all types of samples requires detailed further analysis and adequate sample numbers in each regional or water source cluster.

7.6.1. An MPN design for PA methods

A way to extend the sample range of the comparison would be to subdivide each PA sample after inoculation. Instead of incubating the sample in a single 100 ml test portion, it could be divided equally between 10, 20, 25 or more separate vials for incubation. Microbiologically, the conditions would remain almost the same, but statistically the design would mean a considerable change. Instead of one 100 ml PA test 10, 20, 25 or more parallel PA tests would be made on every sample. The restrictive contamination level of 5/100 ml would be widened. Theoretically, different designs function over the ranges given below.

Design	Theoretical range ((particles/100 ml)
10 × 10 ml	1–23
20 × 5 ml	1–60
25 × 4 ml	1–80
40 × 2.5 ml	1–150
50 × 2 ml	1–200

The added work in the laboratory would be compensated for by the need for fewer samples and the fact that an actual (semi)quantitative estimation of microbe numbers can be computed using the MPN formula for single dilution. The design would also provide better assurance that the number of viable particles in each positive vial is small, only one in most cases, which would be ideal for testing PA methods.

The designs with 20 or 25 parallels suit water studies well. They would be very near the working range of the membrane filter count. An MF count of 100 ml test portions could be used as a reference.

The design is less suitable with solid (food) samples because perfect mixing before sub-sampling is hardly possible. MPN estimation is therefore invalid in cases where microbe numbers are so small that an undiluted sample must be used as inoculum. The parallel sub-samples could only be treated as independent parallel PA tests.

7.7. Quantitative expression and estimation of uncertainty in microbiological methods

7.7.1. *Bias*

Several types of bias have long been known to affect microbiological measurements. The most frequent are:

(i)   dilution bias due to inaccurate volumes (uncalibrated pipettes and dispensers, evaporation of water from dilution blanks in the autoclave or during storage (Lorentz, 1962));

(ii)   crowding (overlap) of colonies on the plate;

(iii)   incomplete recovery of the target population;

(iv)   personal bias in reading colony count results;

(v)   death or multiplication of microbes during sample storage.

Correction factors do not seem to be applied by laboratories, although many of the systematic effects listed above are quantifiable. Evidently, the feeling is that bias is not stable or predictable enough. It is also felt that the imprecision due to random causes is more important. Differences in bias will, however, appear as differences in recovery of micro-organisms with different methods. Thus, even if not specifically measured, bias affects method comparisons implicitly.

7.7.2. *Imprecision: repeatability and reproducibility values*

Repeatability is the random variation remaining in analytical results when all known systematic sources of error have been eliminated. Unknown systematic errors may be included.

For practical use, the repeatability value r and the reproducibility value R have been defined as the values below which the absolute difference between two single test results obtained under repeatability or reproducibility conditions may be expected to lie with a probability of 95%. The calculation of r and R can be derived from the variance of data in an inter-laboratory experiment. The statistical design used for this purpose allows separation of the total variance into two factors, the between-laboratory variance component s_L^2 and the within-laboratory variance component s_W^2. At the 95% confidence level, the values of r and R are then approximated by the following formulae for normally distributed variables:

$$r = 2.8\, s_W$$

$$R = 2.8\, \sqrt{s_W^2 + s_L^2}$$

r and R have been used extensively to characterize the precision of chemical methods for the analysis of water, food, etc. In water and food microbiology, the concepts of r and R have thus far found little use, one of the reasons being the difficulty of preparing homogeneous and stable test materials for inter-laboratory studies. A second aspect is that the data in microbiological experiments are not normally distributed.

Experiences with inter-laboratory tests have shown that the log-normal distribution is most suitable, except with low average counts. r and R back-transformed from the logarithmic to the original scale will represent a critical ratio rather than an absolute difference. A similar approach was presented by Piton and Grappin (1991), who used the geometric relative standard deviation and critical relative difference ($RD = (10^{2.8 s_W} - 1)$) to summarize the results of inter-laboratory tests.

There is an important difference between the application of r and R in chemistry and in microbiology. Because of the discrete nature of micro-organisms and the fact that the count is based on a low number of particles, there is a minimum variation in count results that is described by the Poisson distribution. In this ideal case, the standard deviation of two counts from the same dilution is equal to the square root of the number of colonies counted. In other words, the minimum value of r is related to the average count. Hence there is no single value that can characterize the optimal value of r and R, because it will depend on a factor which is not under the control of the analyst.

The minimum value of r can be derived from the Poisson statistics. It can be shown that r_{min} asymptomatically approaches a lower limit

(Mooijman *et al.*, 1992). In the case of duplicate counts from a single dilution, the limit value is $r = 1.20$. It can also be shown that at an average count of at least 40–50 colonies, the actual value of r_{min} does not differ very much from the limit value, indicating that method performance studies aimed at evaluating r and R should aim at the use of samples which will give a count of at least 40 colonies per plate or membrane.

7.7.3. *Alternatives for estimating the repeatability standard deviation (RSD)*

Estimation of repeatability can be approached in two basically different ways. One is to remove by mathematical means all variation that is attributable to treatments, samples and other recognizable causes of variation. What remains is considered the repeatability variance. This might be called the 'splitting' principle of estimation and it is in use in method comparisons with the aid of the analysis of variance. The other is the 'lumping' principle. Known causes of random error are listed, estimated separately and added together.

If the repeatability estimates determined by the two approaches differ considerably, it is a sign that unrecognized factors are causing variation in the analysis.

(a) The splitting principle

This is the standard practice of estimating repeatability empirically from replicate results on several samples.

Statistical elimination of the between-samples or 'treatments' variation leaves a remainder, the within-sample variation, which is considered the random experimental error. This residue contains all the known and any unknown causes of variation that are not connected with the structure of the experiment. What remains is taken as inevitable random variation.

Sometimes many laboratories participate to obtain a collaborative estimate, which is then assumed to have more general validity. Estimating the repeatability standard deviation in this way produces high values in many cases in microbiology. Data derived from collaborative tests conducted by the AOAC provide illustrative examples (Lancette and Harmon, 1980; Entis, 1986; Ginn *et al.*, 1986; Roth and Bontrager, 1989; Curiale *et al.*, 1989, 1990, 1991; Edberg *et al.*, 1991).

(b) The lumping principle

The idea of estimating the total random error by adding several separately determined error components together was introduced decades ago into microbiological literature. All, or the most significant known sources of random error are estimated separately and compounded by mathematical modelling (Jarvis, 1989) or computer simulation (Dahms, 1992). For the sake of simplicity the components of error combined mathematically are assumed independent. In computer simulation this assumption need not be made.

The principle of combining independent errors is to calculate their geometric sum, or Euclidian distance, as it is also called. To obtain the total error, the squared error components are added and the square root of the sum is taken.

None of the error components are exactly known but estimates have been published. A recent summary can be found in Jarvis

Table 7.3

Magnitude of the major components of random error in a trouble-free colony count method

Error	Symbol	Formula	Magnitude
Counting error	s_Z	–	$\pm 5\%$
Distribution error	s_C	$100/\sqrt{C}$	varies with C
Volume error: 1 ml	s_V	–	$\pm 2\%$[1]
0.1 ml	s_V	–	$\pm 8\%$[1]
Dilution error: per step	s_X	–	$\pm 2\%$[2]
total	s_D	$s_X\,k$	varies with k[3]

1 Depends on volumetric equipment (see, e.g. Jarvis, 1989; Lorenz, 1962).

2 1 ml transfer volume and ratio 1:10 or 1:100 assumed.

3 k = number of dilution steps (transfers).

(1989). Table 7.3 contains what might be considered as reasonable target values for various components of random error in the microbiological quantitative procedure.

The total dilution error given in Table 7.3 is a considerable simplification of the original idea (Hedges, 1967).

The tabulated estimates, or other values if demonstrated to be more representative, can be entered into a mathematical model or be given to a computer simulation programme to produce a lumped estimate of repeatability standard deviation.

The following model includes the most significant random components included in the repeatability standard deviation of the microbiological colony count system. As estimates are used it is customary to substitute the symbol s for σ in statistical formulas.

$$s_r = \sqrt{s_D^2 + s_V^2 + s_C^2 + s_Z^2}$$

where:
s_r = estimate of the repeatability standard deviation
s_D = random dilution error
s_V = random inoculum volume error
s_C = distribution error of colony count C
s_Z = random counting error.

The formula is easiest to apply with relative (per cent) standard deviations. Conversion to logarithmic scale can be approximately effected by division with 230 or 2.30, depending on whether the total error has been computed using percentages or RSD values.

The repeatability standard deviations from collaborative studies are invariably given in $\log_{10}$ scale. They can be converted to ln scale by multiplying with 2.3026.

The standard deviation of the Poisson distribution can be expressed in logarithmic scale by calculating

$$s_{\log} = \frac{0.4343}{\sqrt{C}}$$

where C = the (planned) average colony number per plate, and $s_{\log}$ = standard deviation in $\log_{10}$ scale.

Accordingly, with an optimal number of 80 colonies per plate, the standard deviation under Poisson assumption (distribution error, s_C) would be about 0.05 $\log_{10}$ units. Values obtained by the 'splitting' technique in collaborative studies are usually much higher, frequently greater than 0.20. The truth, presumably, lies somewhere between the extremes.

7.8. Experimental designs for comparing colony count methods

The previous sections have highlighted many factors that affect the analytical results when microbiological methods are applied. Unfortunately it is not yet possible to define a selection of tests and observations that fulfil the minimal or optimal requirements of method validation and implementation. More practical experience is still needed with the recommended procedures.

Selecting the best method among alternatives is a different task. Effective comparisons can be made without ascertaining the absolute validity of any of the alternatives. It only needs to be defined which of the many quantitative performance characteristics is selected as the basis of evaluation. In the majority of cases it is the primary count of colonies.

The mathematical principles of comparing methods do not depend on how much the procedures differ qualitatively. The same statistical tests apply generally, but the choice of the experimental design and the statistical analysis may depend on the nature of the samples. Data transformations may be influenced by the methods being compared.

The analysis of variance is the most versatile statistical tool in method comparisons involving actual numerical data. Its numerous variants can be utilized in detailed studies of method and laboratory performance. With statisticians' support, almost any problem situations can be solved by it.

Some experimental designs and examples of detailed calculations have been published in standards (ISO, 1991). Two designs useful in method comparisons are outlined below.

7.8.1. Method comparisons with unstressed samples or pure cultures

Sample types: Artificially contaminated (spiked) samples and suspensions of pure cultures.

Design: A representative collection of a reasonable number (20 or more) of samples is obtained. One sample at a time is diluted to suitable density of the target organism. When in doubt, several dilutions are examined. The final suspension is either physically divided into as many parts as there are methods to compare, or it is sub-sampled. One determination per method is made on the final suspension (Fig. 7.1).

Limitation: Power of the test is reduced when methods have different relative recoveries in different samples. Moderate to high contamination is necessary except in samples of homogeneous liquids.

Statistical analysis: *t*-test for paired comparisons (for two methods), analysis of variance (ANOVA) for two or more methods.

Data transformation: When methods operate on different starting dilutions (e.g. spiral plating vs. surface spreading, HGMF vs. plating, 'modern' vs. traditional methods), the data must first be transformed to density readings per original sample or per the last common dilution. Otherwise, original colony count data without multiplication by the dilution factor can be used. For final analysis the logarithmic transformation is made.

```
Sample 1 ...dilution..........final suspension    Method 1 —- Measurement
                                                  Method 2 —- Measurement
                                                  Method 3 —- Measurement


Sample 2 ...dilution..........final suspension    Method 1 —- Measurement
                                                  Method 2 —- Measurement
                                                  Method 3 —- Measurement

Sample 3 ... etc.
```

Fig. 7.1. The experimental design in simple comparisons involving two or more methods.

Evaluation: Both statistical and practical considerations are vital. Differences too small to be of practical importance may be found statistically significant. This unlikely occurrence is due to unnecessarily large data. If a mean difference of considerable magnitude is found to be statistically non-significant, more data may need to be collected to find the answer.

Remarks: Splitting or sub-sampling of solid samples ought to be avoided, because a starting suspension with reasonable assurance of random distribution is essential. Samples with relatively high contamination are therefore most appropriate for method comparisons. This may lead to 'censoring' of samples or data. The representativeness of the set of samples and the power of the numerical analysis must be weighed against each other. Low contamination leading to low colony numbers reduces the power of the mathematical test by increasing the random error.

7.8.2. General design to detect inconsistent response to methods or procedures — two-way analysis of variance with replication

The simple tests described above work well only if the differences in relative recovery between the methods are consistent in all samples. In other words, one method is the best in all types of samples. This cannot be generally assumed in natural samples. The relative recovery (or bias) is likely to depend on the physiological state of the target population and on the kind and amount of the non-target population. Both factors may vary from sample to sample, either naturally or by design.

This situation requires an experimental design that includes replication of each method within each sample. The appropriate statistical design is the two-way analysis of variance with replication. Technically, this means parallel plating in the designs presented above.

Field of application: Testing of methods and procedures with target populations suspected of inconsistent reaction to different methods.

Sample types: Samples artificially contaminated with mixtures of pure cultures, naturally contaminated samples, samples with different stresses on target organisms.

	Method 1	Measurement 1 Measurement 2
Sample 1 ......dilution	Method 2	Measurement 1 Measurement 2
	Method 3	Measurement 1 Measurement 2
Sample 2, 3, 4, ... etc.		

Fig. 7.2. The 'basic cell' of the experimental design in a two-way analysis of variance with replication, for comparing methods.

Purpose: Method comparisons, generation of repeatability estimates.

Design: A representative set of samples is obtained. Each sample is diluted to expected suitable density of the target organism. Several dilutions must be examined if in doubt. Each sample can be studied on a different day if necessary. Equal sub-samples of the final dilution are drawn and examined in duplicate with all the methods under consideration (Fig. 7.2).

Limitations: Missing data and zero counts cause problems. Achieving the right dilution is therefore important. Replacing zeros with 1s and missing data with estimated values is not recommended. Instead, a sample with such data should be omitted.

Statistical analysis: Two-way analysis of variance with replication.

Data transformation: In samples where different methods yield countable plates in different dilutions, the counts are first standardized to a common dilution. In other samples, original colony counts are retained. Logarithmic transformation before statistical analysis.

Remarks: If the interaction is not statistically significant it becomes interesting to test the main effects. It may be a matter of interpretation whether the method effects are random or fixed. When testing two batches of the same nutrient medium the effect is undoubtedly random. When the two methods represent different formulations of media it might be acceptable to treat the effects as fixed.

To be on the conservative side, it is best to test the main effect of the methods against the interaction term. The main effect of samples is of no interest under any circumstances because the suspensions are known to have different bacterial densities. An example with detailed computation is presented in the ISO Standard 9998 (ISO, 1991).

Evaluation of method performance on the basis of colony yield alone may not be appropriate with selective methods. An increase in yield may only reflect an increase in false-positive colonies. To get a more complete evaluation, it is advisable to do the same statistical analysis on the confirmed counts and possibly on the confirmation rates as well.

References

Curiale, M.S., Fahey, P., Fox, T.L. and McAllister, J.S., 1989. Dry rehydratable films for enumeration of coliforms and aerobic bacteria in dairy products. Collaborative study. J. Assoc. Off. Anal. Chem., 72: 312–318.

Curiale, M.S., Sons, T., McAllister, J.S., Halsey, B. and Fox, T.L., 1990. Dry rehydratable film for enumeration of total aerobic bacteria in foods. Collaborative study. J. Assoc. Off. Anal. Chem., 73: 242–248.

Curiale, M.S., Sons, T., McIver, D., McAllister, J.S., Halsey, B., Roblee, D.E. and Fox, T.L., 1991. Dry rehydratable film for enumeration of total coliforms and *Escherichia coli* in foods. Collaborative study. J. Assoc. Off. Anal. Chem., 74: 635–648.

Dahms, S., 1992. Simulation als Mittel der Modellkritik: über den Versuch ein Problem mikrobiologischen Arbeitens statistich zu lösen. Diss., Univ. Bielefeld. ISBN 3-89473-479-S.

Edberg, S.C., Allen, M.J. and Smith, D.B., 1991. Defined substrate technology method for rapid and specific simultaneous enumeration of total coliforms and *Escherichia coli* from water. Collaborative study. J. Assoc. Off. Anal. Chem., 74: 526–529.

Entis, P., 1986. Hydrophobic grid membrane filter method for aerobic plate count in foods. Collaborative study. J. Assoc. Off. Anal. Chem., 69: 671–676.

Ginn, R.E., Packard, V.S. and Fox, T.L., 1986. Enumeration of total bacteria and coliforms in milk by dry rehydratable film methods. Collaborative study. J. Assoc. Off. Anal. Chem., 69: 527–531.

Haas, C.N. and Heller, B., 1990. Statistical approaches to monitoring. In: G.A. McFeters (ed.), Drinking Water Microbiology. Springer, New York, pp. 412–427.

Havelaar, A.H., 1993. The place for microbiological monitoring in the production of safe drinking water. In: , G.F. Graun (Ed.), Safety of Drinking Water Disinfection: Balancing Chemical and Microbiogical Risks. ILSI Press, Washington, DC, pp. 127–141.

Havelaar, A.H., 1995. Water analysis — microbiological analysis. In: Encyclopaedia of Analytical Science. Academic Press, London, pp. 5502–5509.

Havelaar, A.H., Heisterkamp, S.H., Hoekstra J.A. and Mooijman, K.A., 1993. Performance characteristics of methods for the bacteriological examination of water. Water Sci. Technol., 27 (3–4): 1–13.

Hedges, A.J., 1967. On the dilution errors involved in estimating bacterial numbers by the plating method. Biometrics, 23: 158–159.

Hernandez J.F., Guibert, J.M., Delattre, J.M., Oger, C., Charriere, C., Hughes, B., Serceau, R. and Sinegre, F., 1991. Evaluation of a miniaturized procedure for enumeration of *Escherichia coli* in sea water, based upon hydrolysis of 4-methylumbelliferyl-β-D-glucuronide. Water Res., 25: 1073–1078.

Hildebrandt, G., Weiss, H. and Hirst, L., 1986. A propos Farmiloe's formula. Fleischwirtschaft, 66: 1128–1130.

ISO, 1991. Water quality — Practices for evaluating and controlling microbiological colony count media used in water quality tests. International Standard ISO 9998:1991(E).

ISO, 1993. Water quality — Vocabulary — Part 8. International Standard ISO 6107-8: 1993.

ISO, 1994. Guide for the determination of repeatability and reproducibility for a standard test method by interlaboratory tests. ISO 5725:1994.

Jarvis, B., 1989. Statistical Aspects of the Microbiological Analysis of Foods. Progress in Industrial Microbiology, Vol. 21. Elsevier, Amsterdam.

Karwoski, M., 1994. Applications of Automated Direct and Indirect Methods for Food Microbiology. Technical Research Centre of Finland (VTT), Espoo, Finland, 121 pp. + app. 69 pp.

Lancette, G.A. and Harmon, S.M., 1980. Enumeration and confirmation of *Bacillus cereus* in foods. Collaborative study. J. Assoc. Off. Anal. Chem., 63: 581–586.

Lightfoot, N.F., Tillett. H. E. and Lee, J.V., 1995. An evaluation of presence/absence tests for coliform organisms and *E. coli*. Public Health Laboratory Service, DOE Contract 7/7/424 Final report, May 1995. Environment Agency, ISBN 011 7532436.

Lorenz, R.J., 1962. Über die experimentellen Fehler bei der Volumenmessung mit Pipetten. Zbl. Bakt. Parasitkde Inf. Hyg., Abt. 1 Orig., 187: 406–416.

Mooijman, K.A., in't Veld, P.H., Hoekstra, J.A., Heisterkamp, S.H., Havelaar, A.H., Notermans, S.H.W., Roberts, D., Griepink, B. and Maier, E., 1992. Development of microbiological reference materials, Commission of the

European Communities, Community Bureau of Reference, Report EUR 14375 EN, ISSN 1018-5593.

Niemelä, S., 1965. Quantitative estimation of bacterial colonies on membrane filters. Ann. Acad. Sci. Fenn., Ser. A VI Biologica: 1–65.

Olsen, J.E., Aboo, S., Hill, W., Notermans, S., Wernars, K., Granum, P.E., Popovic, T., Rasmussen, H.N. and Olsvik, O., 1995. Probes and polymerase chain reaction for detection of food-borne pathogens. Int. J. Food Microbiol., 28: 1–120.

Pitton, C. and Grappin, R., 1991. A model for statistical evaluation of precision parameters of microbiological methods: application to dry rehydratable film methods and IDF reference methods for enumeration of total aerobic mesophilic flora and coliforms in raw milk. J. Assoc. Off. Anal. Chem., 74: 92–97.

Rentenaar, M.F. and Van der Sande, C.A.F.M., 1994. Micro Val, a new and challenging Eureka project. Trends Food Sci. Technol., 5: 131–133.

Roth, J.N. and Bontrager, G.L., 1989. Temperature-independent pectin gel method for coliform determination in dairy products. Collaborative study. J. Assoc. Off. Anal. Chem., 72: 298–302.

Schijven, J.F., Havelaar, A.H. and Bahar, M., 1994. A simple and widely applicable method for preparing homogeneous and stable quality control samples in water microbiology. Appl. Environ. Microbiol., 60: 4160–4162.

Tillett, H.E. and Lightfoot, N.F., 1995. Quality control in environmental microbiology compared with chemistry: what is homogenous and what is random? Water Sci. Tech., 31: 471–477.

Van der Kooij, D. and Veenendaal, H.R., 1992. Assessment of the biofilm formation characteristics of drinking water. In: Proceedings American Water Works Association Water Quality Technology Conference, 15–19 November 1992, Toronto, Canada, pp. 1099–1110.

Van der Kooij, D. and Veenendaal, H.R., 1993. Assessment of the biofilm formation potential of synthetic materials in contact with drinking water during distribution. In: Proceedings American Water Works Association Water Quality Technology Conference, 7–11 November 1993, Miami, FL, pp. 1395–1407.

Youden, W.J. and Steiner, E.H., 1975. Statistical Manual of the AOAC. Association of Official Analytical Chemists, ISBN 0-935584-15-3.

Chapter 8

Analytical quality control in microbiology

8.1. Introduction

Analytical quality control (AQC) is a part of the total quality assurance programme of the laboratory, and should be introduced and evaluated in an integrated manner. The general quality assurance programme will guarantee that all work is carried out according to established protocols (standard operating procedures or SOPs), that equipment and materials are adequately maintained, etc. Analytical quality control is defined as the operational techniques and activities that are used to fulfil requirements for quality (ISO 9000, 3.4).

This chapter describes a practical approach to distributing responsibilities for AQC over different staff-lines in the laboratory, and to defining a time frame for the execution of different aspects of AQC. The principle used is that the checks are differentiated into three lines. The first line is the responsibility of the analyst who actually performs the analyses, and is carried out at a high frequency. The major objectives are to ensure that all aspects of the analysis are under control and that the analyses are consistent over a period of time. The second line is the responsibility of a person independent of the analyst, and is carried out less frequently. The main objective is to ensure that different analysts or pieces of equipment produce similar results or that individual results are not biased. The third line is the responsibility of laboratory management, and is aimed particularly at ensuring inter-laboratory standardization.

The following sections will describe some general aspects of AQC at these three lines, and will give practical examples of tools that can be used in microbiological laboratories. The examples will mainly be derived from (selective) bacteriological culture methods, by enrichment or colony counts, because these methods are highly specific for water and food microbiology.

Quality control of chemical analyses relies to a great extent on accurately known reference materials and standard solutions. It is less easy to produce stable reference materials for quantitative microbiology and the exact number of organisms per sample will not be known. The best that can be achieved for a well-prepared, thoroughly mixed batch is random distribution of organisms which can be described mathematically by Poisson formulae. Such material is sometimes referred to as homogeneous, although in a microbiological context this is not literally true because there is no uniformity in microbial numbers from sample to sample (Tillett and Lightfoot, 1995). Even the best materials available today do not always meet the ideal as predicted by the Poisson distribution.

For this reason, microbiological quality control rests very strongly on comparison of observed and expected or 'acceptable' variability. Internal quality control is best achieved in cases where suspensions or powders are so well mixed that distribution of viable particles can be considered fully random. Whenever two or more readings have been made in such a way that they can be traced back to a common well-mixed sample, the results can be utilized for AQC purposes.

Microbiological viable count methods measure living organisms. The analyte (the microbial population) takes an active part in the quantitative procedure. The analyst, the bacterial population and the culture medium interact in many ways in the process. As a consequence, the results may display variability that is unpredictable but can be understood as a biological phenomenon.

Interactions may be expected to become more frequent and more complex as the colony number per plate increases. The onset of interference may be gradual but it may also reach drastic proportions, particularly with selective media. The former mostly appears in the form of increasing variability with the colony count. The latter causes a sudden loss of conformity of colony counts with sample volume. Both observations contribute to the determination of the upper counting limit of the medium.

A routine monitoring plan frequently does not include duplication of any parts of the procedure. With a single reading per sample, quality control rests on the analyst's ability to detect suspicious growth patterns. Such skills will develop correctly, if at all, unless the analyst has an opportunity occasionally to compare the result with external or internal references. Analysis of the variation of duplicate or replicate determinations in many cases provides valuable QC information.

Some ways of graphical displaying analytical measurement quality are recommended in this chapter. They are based on rather limited experience and can be expected to be refined in the future. Visual inspection can be very informative. The introduction to some charts of statistically expected ranges will begin to explore the usefulness of control charts, which are powerful tools in operational research but are not yet fully adapted and proven with microbiology. In effect, these visual displays are guidance charts used to initiate investigations.

8.2. First-line checks

8.2.1. General principles

First-line checks are carried out and evaluated by the analyst as a means of self-control. Criteria for acceptance or rejection of results should preferably be set beforehand, as well as appropriate action if results do not meet the criteria. In general, the evaluation of first-line checks should be supervised by the direct superior of the analyst, who should also be responsible for setting criteria and defining action plans. It is advisable to carry out first-line checks with every series of analyses. A series is defined as a set of analyses carried out under identical conditions (e.g. same analyst, same batch of culture media and reagents, same apparatus, same time interval, etc.).

(a) Before the analysis

Prior to starting the analysis, it should be verified if all necessary materials and equipment are in a state that conforms to the requirements described in the standard operating procedures. This relates to:

Samples: labelling, sample guidance sheets, storage, preservation and integrity (see Chapter 4).

Equipment: calibration and validation status and eventual adjustment of sample processing equipment, pipettes and other volumetric glassware, incubators, pH-meters, etc. In this context, the following definitions apply (see also Chapters 5 and 6):

(i)   *calibration*: determining the value of deviations of a measuring device as against an applicable standard and, where necessary, determining other measurement properties;

(ii)   *verification*: ascertaining whether the measurement device entirely complies with the rules relating to the nature of the examination and applying at the time of verification;

(iii)   *validation*: testing data against preset criteria;

(iv)   *adjustment*: carrying out the necessary actions to ensure that the measuring device functions sufficiently accurately to be suitable for the intended use.

A simple example illustrates the use of these different concepts. An incubator should maintain a temperature of $37 \pm 1°C$ at all times and at all positions. To *verify* this requirement, a *validation* procedure is carried out. For this purpose, precision thermometers are obtained and *calibrated* against a reference thermometer that is traceable to primary standards. Any deviations are recorded and used to correct the reading of the thermometers. The thermometer(s) is (are) then used to measure the temperature at different locations in the incubator during a certain period of time. If the difference between the lowest and highest recorded temperature is less than 2°C, the incubator can be used for the specified purpose, provided that the mean temperature is 37°C. If the mean temperature deviates, the incubator is *adjusted* by its temperature control knob and the validation repeated.

Culture media, filters and reagents: a laboratory may decide to check certain characteristics before releasing new batches of media, filters or chemicals for use. Such checks may be carried out on bottles of newly bought stocks (e.g. dehydrated media, membrane filters), and may then be valid for the remaining shelf life of the materials. It is also possible to check each lot of ready-for-use materials, e.g., poured plates of culture media, complete reagents, etc. The nature of the materials and the control of the preparation process in the laboratory will determine which action is most appropriate (see also Chapter 6). In any case, the analyst should ensure that the required checks have been made, and that the materials have been released for use. Furthermore, expiry dates, appearance, etc. should be noted.

(b) During the analysis

All information that becomes available during the execution of the analysis should be noted and where necessary recorded. This may indicate if the analysis has been carried out correctly, but will usually only cover certain aspects of the total analysis. Examples of relevant details are:

(i) *General*: temperature registration during sample treatment and incubation, checking anaerobic conditions (if applicable), confirmation rates of typical colonies;

(ii) *Liquid media*: categories of MPN combinations, presence of background flora on plate counts: the presence of background flora, appearance of colonies, spatial distribution of colonies over the surface, drying out of plates or excessive synaeresis water.

(c) In addition to the analysis

To cover all aspects of the analytical process in more detail, additional samples with known characteristics may be included in the series to be analysed. These samples may be prepared in the laboratory, or obtained from another party. Examples are:

(i) *Parallel plating*: a volume from the final test suspension is analysed in duplicate, triplicate;

(ii)   *Procedural blanks*: a sterile liquid is treated by all steps of the analytical process;

(iii)   *Positive control samples*: a stable and well-mixed sample expected to have an average count of the right order of magnitude of a representative strain of the target organism is analysed (such samples may be derived from pure cultures or from naturally contaminated samples, and may be stabilized by freezing or drying);

(iv)   *Negative control samples*: a stable and well-mixed sample expected to have an appropriate concentration of a representative strain of a (mixture of) non-target organisms is analysed;

(v)   *Standard addition*: a positive or negative control sample is added to a sample under investigation and the recovery of the target organism is evaluated; at present, this is possible for presence/absence methods, but no protocol is available for colony counts;

(vi)   *Colony counts on different volumes.*

Several tools for first-line quality control are worked out in detail below. It must be realized that such tools are only beginning to be seriously developed in microbiology, and considerable progress can be expected in the near future. At present, the results of analytical quality control tests should be interpreted with care.

8.2.2. Blanks

To evaluate sterile working conditions throughout the whole analytical procedure, the use of blank samples with every series of analyses is recommended. The common practice of including an uninoculated plate or bottle of growth medium is not sufficient, because it covers only one single aspect of the total analysis. A procedural blank should be used instead, i.e. a sample made of sterile water or diluent that is treated exactly as the samples under investigation, including eventual dilutions, pipetting, filtration, incubating, etc. If possible, the sampling process too should be checked by such methods. If results are consistently good, the programme of sterility checking of glassware and media prior to the analyses can be reduced.

8.2.3. Parallel plating

(a) Background

The distribution of viable particles in a well-mixed suspension is known to follow the Poisson distribution (see introduction to this chapter) although some organisms display natural attraction or repulsion which makes variation slightly larger (i.e. causing 'overdispersion'). For the purposes of these QC checks it will be assumed that such overdispersion is not an influential factor, which assumption has been borne out in the practical experience of the authors. Marked departure from Poisson distribution in a series of parallel colony counts is therefore a sign of problems in the final phase of the quantitative procedure, namely colony development and colony counting.

If parallel plating is not part of the normal practice it should be introduced as a QC measure to a suitable part of the analysis. Selection of samples to this special treatment should follow a predetermined plan, which may be random selection or a systematic plan (e.g. using the nth sample of a batch).

For economic reasons, the number of parallel plates in routine monitoring, if made at all, is not likely to be more than two or three.The compatibility of the observed counts with the expected random distribution of organisms between the parallel portions examined can be tested by calculating either the Poisson index of dispersion (D^2) or its log likelihood equivalent (G^2). See the Annex for calculations.

With two parallels, the root difference is a convenient measure. The QC information it provides is the same as that of the other indices. Its numerical value is close to the square roots of the other two indices (D^2 and G^2).

The statistical yardstick of the two dispersion indices is the chi-square distribution, that of the root difference is obtained from the theoretical constant variance (0.25) of root-transformed Poisson data. The expected standard deviation of the difference is therefore approximately 0.7. Twice that value could be used as the guide value.

Markedly increased variation (overdispersion) is a sign of problems with colony development or with the counting of them. Some of the possible causes, apart from the technician's lack of skill to count colonies, are: contamination, carelessness of the analyst, imprecise volume measurements, interactions between microbial species, malfunctioning of the medium and problems with incubators or other equipment.

Reliable conclusions about medium or analyst performance can be drawn only after cumulating large amounts of dispersion data. Nevertheless, the analysts should calculate and test the dispersion index of each set of replicates, because the chances of identifying the causes of overdispersion are better when the situation is discovered while the plates are still there for inspection.

Identifying the cause of overdispersion is important because the mean based on an overdispersed set may be inaccurate and unhelpful. This is the case when overdispersion is due to contaminated plates or to growth impairment.

(b) Procedure

- Inoculate equal volumes of suspension from the same flask on a set of identical plates (changing the pipette between samples does not make a significant difference).
- Incubate the plates according to the standard practice recommended for the analysis.
- Inspect the plates carefully after incubation and make notes if anything out of the ordinary is seen. Count the colonies.
- Calculate the value of any one of the three homogeneity indices as soon as the colonies have been counted. Refer to statistical tables. In case of substantial overdispersion, check the notes on visual inspection and re-examine the plates if necessary.

(c) Calculations

See the Annex for the mathematical formulae needed to calculate either the Poisson index of dispersion (D^2) or the corresponding log-likelihood statistic (G^2). Mooijman *et al.* (1992) have described the T_1-statistic as a particular application of the D^2-index to examine the dispersion of a large series of parallel plates from different samples. It is therefore particularly useful for application in routine analysis, if parallel plating is carried out. The formula is given in Annex A.

Root difference is the difference of the two values after taking their square roots.

The variation in z between analytical portions from one reconstituted capsule (T_1) and between analytical portions from different reconstituted capsules of a single batch (T_2) were tested separately (Heisterkamp *et al.*, 1992). For the determination of the variation in z between analytical portions of one reconstituted capsule, the T_1 test statistic was applied:

$$T_1 = \sum_i \sum_j [(z_{ij} - z_{i+} / J)^2 / (z_{i+} / J)]$$

where z_{ij} is the result of subsample j of capsule i,

$$z_{i+} = \sum_j z_{ij}$$

the total number concentration of cfps in all samples of one capsule, and J is the number of subsamples of one capsule.

For the determination of the variation between samples of different capsules of one batch, the T_2 test statistic was applied:

$$T_2 = \sum_i [(z_{i+} - z_{++} / I)^2 / (z_{++} / I)]$$

where

$$z_{++} = \sum_i \left(\sum_j z_{ij} \right)$$

is the total number concentration of cfps in all samples of one batch of capsules and I is the number of capsules.

In the case of a Poisson distribution, T_1 and T_2 follow a χ^2-distribution with respectively $i \cdot (j - 1)$ and $(i - 1)$ degrees of freedom. In this case, the expected values of T_1 and T_2 are the same as the number of degrees of freedom. Hence, $T_1/\{i \cdot (j - 1)\}$ and $T_2/(i - 1)$ are expected to be equal to one.

(d) Graphs

Whichever AQC index is selected from the above, its values should be plotted against the $\log_{10}$ mean colony count on the abscissa. The graphs may be easier to interpret if mean counts less than about 10 are analysed separately. On the contrary, counts above the recommended upper counting limit are useful.

It has been seen that the D^2 and G^2 indices occasionally give very high values.These may occur occasionally due to chance or may be indicative of a problem. Such observations should not be censored just because they are extreme. They present a problem when making graphic displays because they may require condensing of the vertical scale. It may be practical to use a square root transformation of scale on the vertical axis to make the chart visually informative.This brings the index of dispersion (D or G) down to the level of the root difference at the same time.

If guidance lines are required on the graph they must be plotted on the same vertical scale, therefore use the square root values of chi-squared if a square root transformation has been used for D or G. Such guidance lines would be called warning and action limits on conventional control charts, but it is prudent not to use this terminology in microbiology since the interpretation will not be as clear and the automatic 'rejection' of a batch of results is unlikely to be the outcome. The guidance lines should be used to check critical factors in the processing and initiate more intense/frequent QC procedures in order to clarify whether there is a problem and identify it.

The dispersion indices (D^2, G^2) have the ability to spot possible underdispersion as well. It is less easy with the root difference. The extreme skewness of the chi-square distribution, even after square root transformation, means that underdispersion is difficult to display graphically by the above control charts. If underdispersion is a concern, it is advisable to plot the dispersion indices together with logarithmic scale on the ordinate. Warning and action limits need to be transformed in the same way.

Sets with different numbers of parallel plates require their own control charts, because the theoretical value of the dispersion index depends on the number of degrees of freedom, which in this case is $n - 1$ (n = number of parallels).

(e) Analysis and interpretation

Values of the chosen homogeneity index are plotted on a control chart against the log colony count. Very little advantage can be expected from a chronological plot. In that sense the graphs are not genuine process control charts. The general principles of Shewart control charts are given in the Annex.

Sample types can be indicated with different symbols.

Even if warning and action limits have been drawn in the chart they should be used more as reference lines than as reasons to stop the entire analytical process. Each individual passing the limit line might be a reason to consider whether the result of that particular analysis should be accepted or not.

An obvious increasing trend of index values with increasing mean colony count may be an indication of an upper working limit of the method. This may be strongly sample-dependent. The sensitivity of the index to additional errors increases with the colony count. A ±10% pipetting error will go unnoticed with a mean of 30 colonies but will appear as considerable overdispersion with 300 colonies. This should be taken into account when analysing the results.

After considerable numbers (100 or more) of index values have been accumulated, their frequency distribution can be compared with the χ^2 distribution with corresponding degrees of freedom. Goodness-of-fit of the two distributions can be tested in the classical manner, but may be unnecessary. Visual comparison of the distributions is likely to provide the relevant information on occurrence of over- or under-dispersion. The use and analysis of dispersion indices is more fully described in the monographs of Eisenhart and Wilson (1943) and the monograph of Stearman (1955).

(f) Example

Three types of food (raw milk, fresh fillets of herring and ground beef) were tested for coliform organisms using the standard violet red bile agar (VRB) overlay method with incubation at 37°C.

After thorough mixing, the samples were diluted in steps of 1:4. Parallel plates were prepared from each dilution. After incubation, the markings were removed and the plates were recoded randomly. Two persons, a trainee and an experienced laboratory worker, counted the number of presumptive coliform colonies on each countable plate according to the description in the standard IDF 73A:1985. The recommended counting range of 15–150 was not obeyed.

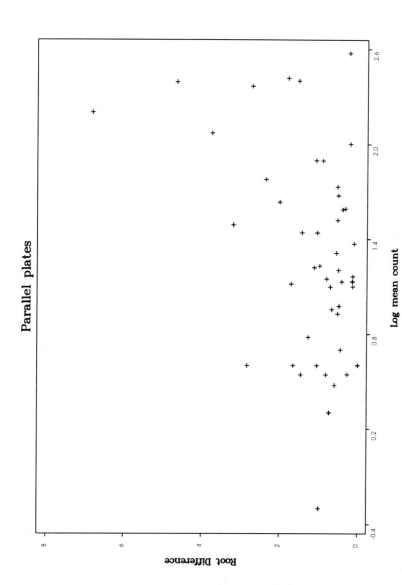

Fig. 8.1. Fit of colony counts on parallel VRB plates. Data on milk, herring, and ground beef combined. RD = root difference, abscissa: $\log_{10}$ mean count per plate.

Data of the more experienced reference person were analysed for homogeneity of the parallel counts. The absolute root difference $\text{RD} = \left| \sqrt{C_1} - \sqrt{C_2} \right|$ was used as the index of agreement between the parallels (see Fig. 8.1).

The results show a rising trend. If values of root difference twice the standard deviation (i.e. about 1.5) are considered satisfactory, it is seen that the fit is good only up to about 40 colonies per plate (log mean count = 1.6). This gives an indication that the VRB overlay method may not really have the working range of 15–150 colonies per plate that the standards assume.

8.2.4. Colony counts on different volumes/dilutions

(a) Background

Because of the great variability in bacterial concentrations in natural samples, and the limitations on the number of colonies counted on a single plate, it is normal practice to examine two or more different volumes (dilutions) of every sample. It is customary to use ten-fold ratios between successive portions. Smaller ratios are often more convenient with some techniques, such as membrane filtration. They are also more appropriate for QC purposes.

Frequently, more than one sample portion yields countable numbers. The common convention of considering reliable only those counts between 30 and 300 (or 25 and 250, or 15 and 150), may eliminate all but one of the counts from consideration. Selecting one 'reliable' number may be sound practice as regards reporting the bacterial content of the sample.

There are different schools of thought as to how to assign the count per unit volume of the sample, when counts on two or more volumes of the same sample are available. Indeed, somewhat arbitrary decisions based on common sense may be inevitable.

Even if only one count is selected for calculating the sample bacterial content, counts on other sample portions, also outside the given limits, should be recorded. They may provide helpful information on the performance characteristics of the method.

Consider a pair of colony numbers (C_1 and C_2) derived from the cultivation of the respective volumes V_1 and V_2 ($V_1 > V_2$). It is to be expected, on average, that the count ratio C_1/C_2 will be equal to the volume ratio $a = V_1/V_2$. Significant departure from the expectation indicates problems with the quantitative procedure.

A statistical test of agreement between the two counts can be based on either of the distribution indices discussed in the Annex. Overdispersion indicates departure from the expected ratio. As volume errors are unlikely to be a sufficient explanation, such observations are believed to indicate problems with colony development on one of the plates.

It is generally assumed that high colony numbers are more likely to be affected by biological interactions and other problems than moderate ones. It is to be expected that the count ratio will cease to agree with the volume ratio when the higher of the two colony numbers exceeds a limit, where the crowding of the bacterial population breaks down the selectivity or specificity of the medium, or where the technician cannot manage the counting. This should become apparent as an increase in the value of the dispersion index. ASTM has based a test of the upper counting range of a medium on that idea (Dufour, 1980).

The statistical test is rather 'conservative' with high volume ratios. Data with ten-fold ratios may have to be collected over a long time before the upper counting limit emerges through weight of evidence. Graphs where values of the dispersion statistic are plotted against the expected count $E(C_1) = aC_2$ (a = volume ratio: V_1/V_2, C_2 = the colony number in the lower volume) can be utilized for the purpose.

Statistically, the test is most powerful when both the colony numbers C_1 and C_2 are high. Decreasing the volume ratio will accordingly increase the statistical power of the test. However, detection of the actual microbiological problem may require that C_1 and C_2 are not very close to each other. As a compromise, volume ratios between 3:1 AND 5:1 are recommended.

A way to increase the power of the test is to make parallel plates. ASTM prescribes three parallels for both volumes. Three is a suitable number if the working limit of the method is in the vicinity of 300 colonies per plate. With lower expected limits the power of a single test may not be sufficient unless more parallels than three are made.

(b) Calculations

(i) One plate per volume
Let:
C_1 = colony number in volume V_1
C_2 = colony number in volume V_2
a = V_1/V_2
Calculate one of the following indices:

$$D^2 = \frac{(C_1 - aC_2)^2}{a(C_1 + C_2)}$$

$$G^2 = 2\left(C_1 \ln \frac{C_1}{a} + C_2 \ln C_2\right) - 2(C_1 + C_2)\ln\left(\frac{C_1 + C_2}{a + 1}\right)$$

Additionally, calculate the quantity $A = \ln(C_1 V_2/C_2 V_1)$ (Gameson, 1983) ignoring data where C_1 or C_2 equals zero.
In cases of parallel plating calculations change as follows.

(ii) Equal numbers of parallels in V_1 and V_2
Let:
C_1 = sum of colony numbers in volumes V_1
C_2 = sum of colony numbers in volumes V_2
Then proceed as above.

(c) Graphs

Plot the square root values of D^2 or G^2 (i.e. D or G) on a control chart with the logarithms of the expected higher colony numbers per plate on the abscissa. The expected count in different cases equals:
(i) One plate per volume. Expected higher count $E(C_1) = aC_2$,
(ii) Parallel plates. Expected mean higher count $E(C_1) = aC_2/n$, where C_2 = sum of colony numbers in smaller volumes (V_2), n = number of parallels.
(The reason for choosing the expected count instead of the mean count observed in the higher volume (V_1) is that the higher colony count is the one likely to be affected, and may have become a totally erroneous one.)

(d) Analysis and interpretation

Visual inspection of the chart should suffice. Guide levels derived from the chi-square distribution with one degree of freedom may be added to support conclusions and introduce some parallel with control charts.

Alternatively, the data can be arranged in a table according to increasing values of the higher expected count highlighting cases where significant overdispersion occurs. High frequency of overdispersion at high colony densities signals the upper counting range of the medium.

The direction of discrepancy is not apparent in the G^2 and D^2 values. Depression of C_1 in comparison with aC_2 is an indication of biological interactions on the plate. Gameson (1983) has presented observations of the opposite nature on selective coliform media, the count ratio tending to exceed the volume ratio when coliforms were numerous. This happened because non-coliforms were identified as coliforms.

The data can be further inspected by plotting the values of A on a control chart. Negative values indicate that $C_1 < aC_2$.

Great scatter of index values at low colony numbers is only natural because of the nature of the Poisson distribution.

(e) Example

Refer to the example given in Chapter 8, Section 2.3. Data of the more experienced analyst were analysed for linearity of colony numbers vs. dilution. Data sets where the total colony number of the higher dilution was less than five were omitted.

The square root of the dispersion index (D^2) was used as the AQC measure (Fig. 8.3). If we take a conservative attitude and regard the 1% point of the dispersion index as the limit of a reasonable fit, values of D beyond about 2.5 to 2.6 indicate problems. This limit was approached already when the expected colony count per plate was about 40 (log expected mean $C_1 = 1.6$). The conclusion is remarkably similar as with the fit of parallels (see Fig. 8.2).

Thus, in this example the linearity of counts with dilution began to get 'out of control' when 40 colonies were expected per plate. This was rather unexpected, because the standard method is assumed to

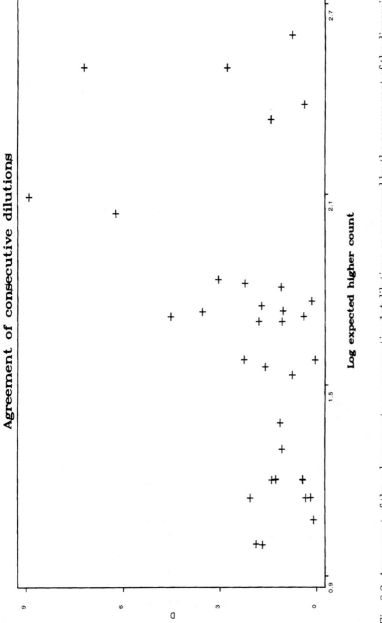

Fig. 8.2. Agreement of the colony counts on consecutive 1:4 dilutions as measured by the square root of the dispersion index (D). Abscissa: the logarithmic expected mean colony count per plate of the lower dilution. Cases with lower counts $C_2 < 5$ were omitted.

function up to 150 colonies per plate (log 150 = 2.18). It may not be possible to find a single cause for this. It is unlikely that technical volumetric errors would account for the discrepancy. Rather the nutrient medium or the analyst was not quite able to handle the bacterial population of some sample types.

These examples illustrate only one case. The same QC indices may behave quite differently in other cases, leading to different interpretations. There is no theoretical reason to assume that lack of agreement between parallel plates and lack of fit of dilutions should always occur at the same colony density.

8.2.5. Quantitative quality control samples for first-level quality control (pour-plates, spread-plates, membrane filtration)

First-level quality control samples should be easy to prepare, easy to use, stable, homogeneous and clearly respond to critical aspects of the QC check. Basically, two types of samples can be distinguished: pure cultures and naturally contaminated samples. Naturally contaminated samples may be more realistic, but are more difficult to prepare with sufficient homogeneity and reproducibility. It is advisable to set up a QC programme using pure cultures and to supplement with naturally polluted samples at a later stage. The samples need to be stable in order to get meaningful results and to be able to compare the analytical performance over a period of time. The required time period depends somewhat on the nature and size of the laboratory, but in general it can be said that in order to pay off the time and costs invested in preparing and characterizing the samples, a stability of at least half a year is required. Ideally, QC samples should be stable for at least a year and it should be possible to make them in large batches.

Stabilization of QC samples can be carried out by freezing or by drying. Other methods do not give sufficient stability. Stabilization by drying can be carried out through the freeze-drying process or the spray-drying process. Both methods require considerable skill and experience before they can be carried out reproducibly. It is not advisable for individual laboratories to prepare their QC samples by these processes. Dried materials can be obtained from central laboratories or from commercial suppliers. When evaluating the available materials, special attention should be given to the concentration of test bacteria and their stability (long-term storage under laboratory conditions and short-term effects of transport at ambient temperatures) as well as the homogeneity of the samples. There is theoretical evidence that in order to obtain efficient statistical power, QC samples for colony counts (pour-plates, spread-plates, membrane filtration) should give 40–80 colony forming particles in an analytical portion. QC samples for qualitative (presence/absence) tests should contain about five colony forming particles per analytical portion. It cannot always be expected that the counts between different samples follow the Poisson distribution, but the overdispersion of the samples should be known and within certain limits. The narrower these limits, the better the material. Obviously, the price and the ease of use are also important factors. Practical QC samples, prepared from spray-dried artificially contaminated milk have been developed under the auspices of the European Commission by the Netherlands' National Institute of Public Health and Environmental Protection. Full information can be found in Mooijman *et al.* (1992), and further information can be obtained from: SVM, P.O. Box 457, NL-3720 AL Bilthoven, the Netherlands.

Preparation of QC samples by freezing is more suitable for use by individual laboratories. It has the advantage that it is more flexible (i.e. also applicable to strains that do not survive drying well, or would give safety problems; and the contamination level can easily be varied). On the other hand, preparing and checking frozen samples requires more expertise and costs more time than buying available dried samples, which may be particularly difficult for smaller laboratories. Frozen samples can be transported to other laboratories, but at a greater expense than dried samples. The method described below for preparing frozen samples gives minimum cell damage and guarantees maximum stability and reproducibility. It has been tested extensively with a variety of bacteria (Schijven *et al.*, 1994). The method uses rapid freezing of a suspension of bacteria in skimmed milk in ethanol dry ice or liquid nitrogen, and rapid thawing in a 37°C water bath to induce minimum change in the conformation of cell membranes and cell contents. This is particularly critical for vegetative bacterial cells, but less so for bacterial spores, bacteriophages, etc. Test samples from these organisms could be prepared simply by placing in a −70°C deep freezer, and thawing at room temperature It has been shown that these samples were homogeneous and stable for at least a year. Any laboratory introducing the use of these QC samples should verify the homogeneity and stability of the preparations before use, by following the instructions provided with each package.

8.2.6.   Use of control charts

In industry (operational research) with many quantitative laboratory exercises, an important tool in routine quality control is the control chart. In its basic form it is constructed by drawing horizontal lines for warning and action limits symmetrically above and below the expected mean result. The spacings are derived from estimates of standard deviations of normally distributed data. Control observations are usually plotted sequentially on the chart in order to detect drifts in product quality. They can lead to the discarding of a batch of products in an industrial situation. In microbiology, control charts are being considered and introduced with caution, they are not used to discard batches of results but are in effect guidance charts which are used to trigger investigations. The use of control charts is presently recommended mainly as a tool for continuous improvement, rather than a vigorous check on the variability of analytical data.

Some of the most useful control variables in microbiology (e.g. D^2 and G^2) are based on the asymmetrical chi-square distribution. Symmetrical warning and action limits cannot be constructed. Transformations of data are helpful towards producing more nearly symmetrical limits. Ideally, the data would follow a Poisson distribution, and square root transformation would be most appropriate. For practical purposes, a $\log_{10}$ transformation is preferred, because in most cases overdispersion is unavoidable, for example because samples are examined on different days using different batches of media. As most laboratory technicians prefer to work with untransformed data, the results on the $\log_{10}$ scale can easily be back-transformed to construct the final chart. The back-transformed chart is asymmetrical. Another workable alternative is to construct the chart on graph paper with logarithmic scale on the ordinate.

In most cases, control data plotted with the colony counts on the horizontal axis produce additional information compared with charts in temporal order. At this stage of development of microbiological AQC it may be most appropriate to consider control charts as convenient graphical methods of visualizing method and analyst performance. Warning and action limits drawn from statistical distributions may be used as points of reference, but hardly as reasons for stopping the machinery if exceeded. There is too much uncertainty about the actual underlying distributions.

The following example gives a method for constructing a control chart for use with quantitative QC samples as described in Section 8.2.5.

The use of control charts consists of the following steps:

(i)　Prepare and evaluate a batch of a stable and well-mixed quality control sample (see Section 8.2.5);

(ii)　Examine 20 aliquots of the test sample, preferably on different days and involving all personnel and equipment that will normally be involved in the analysis. (It is possible to construct a control chart from 10 measurements; the calculations should then be repeated as soon as 20 observations have been made.);

(iii)　Calculate (on the $\log_{10}$ scale) the mean (x) and the standard deviation (s). For microbiological counts, a robust estimate of s is preferred above the usual formula. Such a robust estimate is likely to be only slightly affected (in contrast to the usual formula) by variations in counts due to assignable causes. Calculate as follows:

$$\overline{R} = \frac{1}{n-1} \sum_{i-2}^{n} \left| x_i - x_{i-1} \right|$$

$$s = 0.8865 \times \overline{R}$$

where n is the number of observations and x_i is the ith observation.

From these results, construct the control chart. The control limits are (on the $\log_{10}$ scale):

(a) warning limit $x \pm 2s$,
(b) action limits $x \pm 3s$.
Back-transform the mean and control limits to the original scale, and draw the control chart.

(iv)   Analyse an aliquot of the test material with each series of analyses (preferably as the last sample), plot the results on the chart.

(v)   Check the results against the control limits and act accordingly; the results may be out of control, if

  – there is a single violation of the action limit;
  – two out of three consecutive observations exceed the same warning limit;
  – there are nine consecutive observations on the same side of x;
  – six consecutive observations show a continuous rising or decreasing trend.

    If results are outside these criteria, the reason for the erroneous result should be identified. In the present stage of development of control charts for microbiological analyses, it is not possible to directly link a result which is out of control to the validity of the results obtained. A decision about the validity of the results of the particular series should be taken by laboratory management on the basis of the identified reasons for the deviating result. The decision-making process should include critical assessment of the statistical analysis alongside detailed scrutiny of all other evidence, such as records of equipment, temperature, etc. The use of control charts is presently recommended mainly as a tool for continuous improvement, rather than a rigorous check on the validity of analytical data.

(vi)   Construct a new control chart if necessary, as long as the test material is available and stable.

    An example of the construction and use of a control chart is given in Table 8.1 and Figs. 8.3 and 8.4. The charts were prepared using a frozen suspension of a test strain, which responds very well to changes in the quality of bile–esculin

Table 8.1

Example of the calculations of limits for control charts. KF–Streptococcus agar and bile–esculin–azide agar. Test strain WR78.

Test no.	KFA cfp	BEAA cfp	KFA log-transformed	BEAA
1	36	26	1.556	1.415
2	32	29	1.505	1.462
3	43	47	1.633	1.672
4	40	39	1.602	1.591
5	42	31	1.623	1.491
6	49	30	1.690	1.477
7	29	44	1.462	1.643
8	41	57	1.613	1.756
9	39	56	1.591	1.748
10	36	38	1.556	1.580
11	32	43	1.505	1.633
12	44	50	1.643	1.699
13	40	15	1.602	1.176
14	37	31	1.568	1.491
15	46	39	1.663	1.591
16	36	33	1.556	1.519
17	43	34	1.633	1.531
18	43	39	1.633	1.591
19	41	37	1.613	1.568
20	32	31	1.505	1.491
21	48	31		
22	42	40		
23	44	31		
24	39	39		
25	40	36		
26	32	10		
27	49	21		
28	45	32		
29	43	39		
30	28	49		
31	44	37		
32	37	49		
33	41	29		
34	39			
35	36			

(*continued*)

Table 8.1 (*continuation*)

	KFA (log-transformed)	BEAA
Calculations on first ten measurements		
Mean	1.583	1.584
Standard deviation robust	0.072	0.089
Control limits robust after backtransformation		
upper action	63.1	71.1
upper warning	53.4	57.8
mean	38.3	38.3
lower warning	27.5	25.4
lower action	23.3	20.7
Calculations on first twenty measurements		
Mean	1.588	1.556
Standard deviation robust	0.066	0.103
Control limits robust after backtransformation		
upper action	60.9	73.4
upper warning	52.3	57.9
mean	38.7	36.0
lower warning	28.6	22.4
lower action	24.6	17.7

agar, as can be seen in the example. The data on KFA show only random variation, and no results are out of control. The data on BEA were produced in parallel to those on KFA, and two results (#13 and #26) are outside the lower warning limit. This indicates that on these days, the preparation of the BEA medium obviously had some defects. Note that the inclusion of the deviating measurement #13 indeed affects the control limits only slightly. If a technical reason for this outlying result were detected, it is advisable to recalculate the control limits excluding this single result.

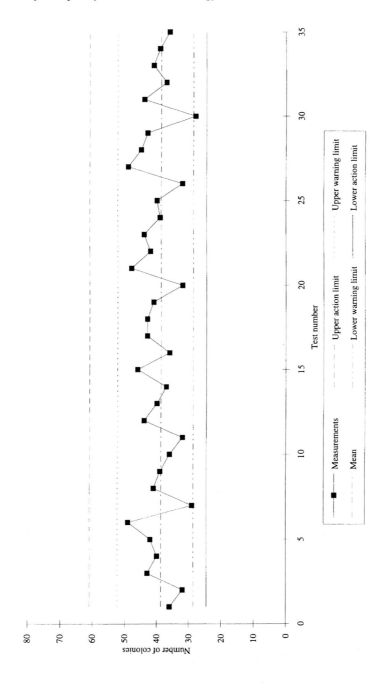

Fig. 8.3. Control chart of *E. faecium* WR78 on KFA.

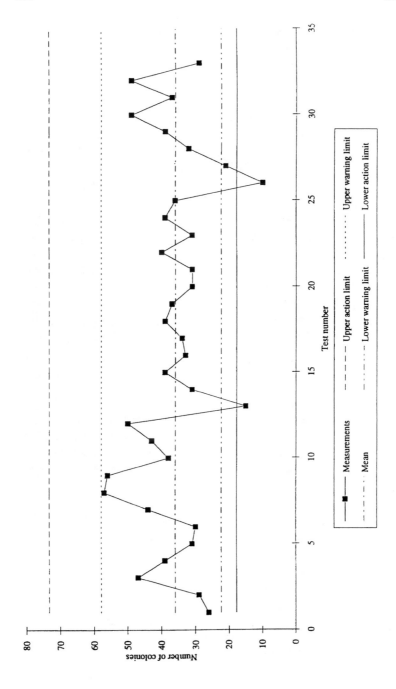

Fig. 8.4. Control chart of *E. faecium* WR78 on BEA.

8.2.7. Control of the MPN technique

The basic assumption in MPN estimates is that the bacterial density in any dilution or sub-sample is proportional to the main sample density. This assumption is violated if bacteria have unequal opportunities for development in different sub-samples or if sub-sample volumes are not what they are believed to be. Quality control of the MPN technique aims at detecting violations of the basic assumption.

The traditional MPN technique with three or five parallel tubes per dilution is a crude instrument. Inhibition of the growth of the target organisms and gross volume errors cause irregularities that appear as 'improbable' series of positive tubes. Any ordinary inaccuracies in volume measurements are not likely to be detected. Due to the great random variation under the basic model, simple split sample quality control is ineffective.

A recent computer-based MPN data-handling programme incorporates a test of data quality (Hurley and Roscoe, 1983). Older publications formulate rules and tests for detecting improbable series of positives that can be applied without computers (Taylor, 1962; Sen, 1964; de Man, 1975, 1983).

(a) Deviance test

Hurley and Roscoe (1983) formulated equations for determining the MPN estimate as well as its standard error and confidence interval for a general case with any number of parallels and any dilution ratios. In addition, they provide a statistical test of homogeneity based on the 'deviance'. The somewhat involved calculations can be automated by the use of a BASIC computer programme provided. The test statistic D (for 'deviance') is approximately chi-square distributed with $k - 1$ degrees of freedom (k = number of dilution levels).

(b) De Man's categories

De Man's categories can be used whenever the MPN design is based on dilution ratio 10 and three or five parallels per dilution. De Man (1975, 1983) divided the possible series of positive responses into categories according to how probable each one is under the basic model. It is recommended that only MPN series of the first and second category be accepted as satisfactory.

(c) Elimination of bias

The bias of the MPN estimate has been discussed and debated for a long time. It seems well established that the MPN function remains skew even after logarithmic transformation. The MPN estimate is a biased estimate of the mean. A computer programme developed by Klee (1993) yields estimates corrected for bias. A corrected table is also given. Unbiased MPNs for selected combinations of dilution combinations have been calculated from exact conditional probabilities (Tillet and Coleman, 1985). The magnitude of bias is usually small, and negligible compared with sampling variations (Tillett, 1995). Therefore it would seem unnecessary to correct for bias but it is advisable always to use the same tables or computer program to estimate the MPN for comparability.

8.2.8. Confirmatory tests

Most selective methods are procedures with more than one stage. With P/A methods the successive stages are an integral part of the process and they are always carried out fully. With selective colony count methods, the standard protocols usually regard confirmation as something to be done 'when necessary'. A few colonies are picked for confirmation. The numbers typically recommended are too small to serve quality control purposes. (Besides, precision of the estimates is spoiled by the small isolation rate.)

In particular, laboratories that tend to trust the primary colony count as a valid final result should check rather frequently the validity of this trust by randomly isolating presumptive target colonies. The number isolated must be adequate — at least 30 to 50 per sample. More parallel primary plates are necessary in these special experiments than is customary in the routine. All the major sample types should be covered in the long run.

Reagents and media used in confirmation tests should be controlled with the aid of positive and negative cultures.

It can arbitrarily be recommended that samples with confirmation rates constantly higher than 80% do not require confirmation in the routine. A desirable outcome of the special verification tests is a list of samples or sample types that do not require confirmation. Seasonal variation of bacterial populations in natural waters may also mean that verification is necessary during certain periods (snow-melt, rainy seasons) but not in others.

The confirmation rate (number confirmed/number isolated) is a QC variable of considerable value. Its standard deviation depends on both the number of colonies isolated and the number confirmed, and is therefore not fixed. Conventional control charts cannot be applied. A convenient visual check can be based on graphs where each confirmation rate observation is plotted with its standard deviation shown with a line segment.

It should be kept in mind that confirmation rate is not a stable method characteristic. Its value depends on the microbial target and non-target population, and therefore on the sample material and even seasons in environmental samples. It has a personal element, too, because different analysts interpret the colony picture differently. Comparison of the confirmation rate of different analysts or laboratories can therefore be used to judge method robustness (see Section 7.4).

The standard deviation of the confirmation rate is calculated according to the binomial distribution:

$$s_p = \sqrt{\frac{p(1-p)}{n}}$$

where p = confirmation rate (k/n), n = number isolated, and k = number confirmed.

Sample types and analysts should be indicated by distinct graphic symbols if separate graphs are not made. A time scale which actually indicates the time of the year, not only temporal sequence, should prove useful on the abscissa.

8.3. Second-line checks

8.3.1. General principles

Second-line checks are carried out periodically by a person independent of the analyst. This may be a technical manager, or the quality assurance manager. This person also evaluates the results. The process is supervised by laboratory management. The major aim of second-line checks is to assure reproducibility between different analysts or different pieces of equipment. This is particularly important in microbiology, where subjective interpretation of cultures, microscopic preparations, gels, etc. is common to most methods. Laboratory management must assure that new workers maintain the same standards as their experienced colleagues, and also that these standards are maintained at a constant level over time. Therefore, second-line checks are important to evaluate the success of a training period, as well as to regularly evaluate resident staff. Some methods are highly sensitive to different interpretation, whereas other methods are not. It is hoped that widespread use of second-line checks will produce data on the ruggedness of methods in this respect, thus enabling a better selection of methods for national or international standardization.

8.3.2. Duplicate counting

(a) Background

Counting the colonies of a plate more than once provides data for calculating the counting 'error'. Normally, an analyst repeats the counting only when he or she suspects a mistake in the first counting, and seeks the advice of a colleague only in problematic situations. Such sporadic and selective checks do not supply proper quality control information.

Duplicate counting applied in a systematic way is a rather powerful quality control tool. A suitable portion of the plates should be randomly allocated to repeated counting. This should be done in such a way that the analyst does not know during the first counting whether the plate is to be counted again. This ideal scheme may be too impractical to carry out in the routine. In that case, it ought to be considered if a periodical, more concentrated, action would be more appropriate. The examples considered at the end of the chapter illustrate the latter approach. In any case, the work should be distributed over a long time in order to include a large range of different samples and colony densities.

When the counting is repeated by the same person, there is no systematic component. The pairs of counts provide the counting error under repeatability conditions. A basic problem in this connection is how to avoid the first count affecting the second.

When different persons participate, the variation includes both random and systematic components. The data can be used for studying personal differences and for training purposes. If all participants are experienced analysts, the data provide material for calculating a consensus counting 'error'.

Before any generalized statements are made, the data should be inspected graphically. In a training situation it might seem natural to plot the differences sequentially in order to follow the learning process. In all other situations it is more informative to plot the values against colony numbers in order to draw conclusions about the upper and lower counting range or about the technicians' relative ability to cope with the increasing complexity of the biological interactions.

(b) Procedure

The plates destined for recounting should be selected one by one during the routine. Any suitable method of randomization (playing cards, coloured or numbered balls, random sampling numbers, etc.) can be used.

If it is feared that the knowledge of the previous result too much affects the second counting the plate may be set aside to wait for other plates, provided that no more than an hour passes between the first and second counting. The time limit can be somewhat extended by storing the chosen plates in a refrigerator. This is necessary, especially when several laboratories are involved. Any markings must be removed before the second counting.

(c) Computations

Counting errors and biases are expected to be proportional to the colony number. The most useful QC variable for duplicate counting is therefore the relative difference (RD), which can be calculated in two ways. The relative difference (RD) and logarithmic difference between two quantities are namely numerically roughly equal. Therefore, either of the following quantities is equally suitable:

$$\text{RD} = \frac{C_1 - C_2}{\overline{C}} \simeq \ln C_1 - \ln C_2$$

where C_1 = first count, C_2 = second count, and C = average.

Both can be expressed as percentages through multiplication by 100 if that seems more convenient.

When the two counts are by the same person the absolute value (value without sign) of RD is the most appropriate. When the counts are by different persons, the sign also is important.

When more than two analysts participate, pairwise signed differences can be calculated taking one of the persons as reference, if appropriate. If the group consists of equal experts the standard deviation may be a more relevant control variable. It is calculated according to the formula given in the Annex.

The standard deviation and the relative difference are not the same. With two values, the relative standard deviation (RSD) is equivalent to the relative difference divided by the square root of 2. It is not always entirely clear which expression different authors have in mind when writing about the counting error.

(d) Graphs

Usually, no-one can claim to have the right answer in colony counts. Selection of the reference person among a group of experts is therefore arbitrary. In a training situation, one of the persons can be termed 'the expert' and be taken as the reference.

The values of the appropriate QC variable (signed or unsigned RD or standard deviation) should always be plotted against the colony number per plate. In training, another control chart can be based on QC values plotted in chronological order.

The colony count on the abscissa can be that of the reference person or the average of all participants. The low and high ends of the colony count range are the most important areas for drawing conclusions. To fit both ends in the same graph in sufficient detail, logarithmic scale on the horizontal axis is the most appropriate. In this case, logarithms to base 10 are recommended because of their familiarity.

There is no theoretical guide value for the permissible difference between the counts on the same plate by the same or different persons. Obviously, there is a (hidden) true number of colonies on the plate. Ideally, the count should be exactly reproducible but in practice some inaccuracy must be permitted. Experience shows that differences (RD) up to 5 or 10% are 'normal'. Quoting the empirical findings of Fowler *et al.* (1978), the quality control handbook of the Nordic Committee on food analysis requires that an analyst should be able to repeat his or her own count with an accuracy of 7.7% but permits a difference of 18.2% between different persons. Considering the simplicity of the task, it is rather doubtful that such a large difference should be acceptable between different analysts.

(e) Analysis and interpretation

Theoretically, the variation should have no trend over the whole useful counting range, except that in the main no differences would be expected with colony counts of less than about 25.

Considering the simplicity of the counting process, the result should not be substantially different when more than one person participate. The fact that larger variations are normally observed indicates systematic personal differences in counting ability. Where to set the limit of acceptable personal differences in interpretation has not been generally agreed. Very large differences are clear indications of the lack of robustness of the method. This in turn may have its roots in insufficiently detailed descriptions of target colonies. Increasing variation in the high colony count range indicates disturbing medium-analyst-microbial population interactions. The result indicates that the method does not function properly with the type of sample studied, because different analysts get different results.

A marked increase of variation at the low colony count range may occur with selective media. This indicates differences between analysts in the interpretation of colony morphology.

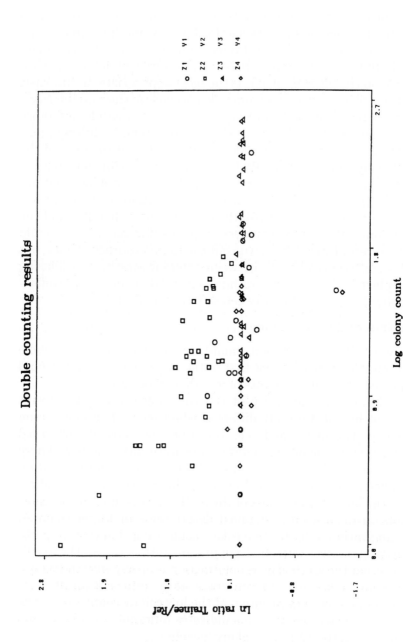

Fig. 8.5. Results on counting of colonies by two persons. VRB agar 37°C. A = trainee, R = reference person (experienced technician). $Y = \ln(A/R)$, $X = \log R$. $\bigcirc$ = milk, $\square$ = herring, $\triangle$ and $\Diamond$ = ground beef.

(f) Example

The results of double counting may reveal important facts not only concerning the analysts but also the methods and sample materials.

When two persons count colonies in an unproblematic situation, where the target colonies are obvious, they should get counts that differ by no more than about 5% (± 0.05 ln units or $\pm 0.02 \log_{10}$ units).

Figure 8.5 shows a case where the interpretation of the trainee differs very much from that of the reference person in one type of sample (fish). The differences are so great that the situation cannot simply be ignored. In the other samples, the differences are more or less within acceptable bounds.

The very clear negative trend provides a probable explanation. There must have been at least two not very clearly distinct colony types in the fish sample (coliforms and *Aeromonas*, presumably). The trainee counted them both while the reference person counted only one type. The two colony types differed most distinctly when there were only very few of them on the plates. As colony numbers increased they became less and less clearly distinguishable, and after about 80 colonies per plate neither analyst was able to tell them apart. Both were getting similar results.

Should one conclude that the method is unsuitable for fish samples, unsuitable in general, or that more training is indicated? The answer may depend on additional observations.

8.3.3. Duplicate analytical procedure

(a) Background

To test the whole quantitative procedure, including dilution, duplicate samples must be studied. This is not a standard practice in most laboratories, but should be considered regularly to keep the procedure under control.

The classical type of control charts as used outside microbiology, for example Shewart charts with values plotted chronologically, may be appropriate. To obtain the standard deviations required for their construction, the alternatives discussed in Section 7.3 can be considered. See the Annex for general principles.

(i)   The minimal (Poisson) estimate of repeatability standard deviation is appropriate when dilution is not involved.

(ii)   When dilutions are involved the standard deviation can be estimated by mathematical modelling, computer simulation or empirically. The first two alternatives are preferable (see example in Section 7.3.)

It is imperative that a 'homogenous' (i.e. microbiologically well-mixed) suspension has been attained by the stage where replication of the dilution series begins whenever the data are used for QC purposes.

(b) Procedures

Mathematically different situations can be recognized. In all cases the results should be read in ignorance of the other result, if possible.

(1) Binomial case
In the hygienic water control, the sample plated for colony counting may consume most or all of an undiluted field sample. 100 ml filtered for coliform counting is an example. To simulate this process for internal QC purposes, a 200 ml sample is divided after thorough mixing into two 100 ml portions, which are examined *in toto* by the membrane filtration technique.
The two samples are put through the analytical procedure as separate samples.
In this case there is no Poisson process. Control chart guide values are based on the binomial distribution as presented by Tillett and Lightfoot (1993).

(2) Poisson case
Some laboratories make routine milk analyses by plating directly a small ($1\ \mu$l) sub-sample of undiluted raw milk. Blind duplicate samples are introduced. Accurate volume measurement is not critical. Pass through the analytical procedure as separate samples. The inoculum should be a small undiluted fraction of the sample.
In this case, colony counts ideally follow the Poisson distribution. Data from a split sample can be treated like parallel samples from a

single 'homogeneous' suspension as was described in Section 8.2.3.

The knowledge that the variance of a root-transformed Poisson variable is nearly constant (0.25) can be used for constructing a conventional control chart.

(3) Dilution case

Samples with a high bacterial content need to be diluted to suitable particle density before plating. The dilution series is basically a chain of Poisson processes (with a high sampling fraction, however). This increases the variation. Furthermore, each volume measurement in the dilution series adds an element of random error to the final count. For these reasons it is inevitable that the colony numbers at the end of independent dilution series vary more than the simple Poisson model predicts.

A check on this technical procedure consists in making two (or more) independent dilution series from the first homogeneous suspension and determining the particle density of the final suspensions by colony counting.

Statistical control of the procedure requires an estimate of the repeatability standard deviation on which to base a control chart. Interestingly, there are several ways to approach this question in microbiology. They range from the minimal single-step Poisson estimates through mathematical models (Jennison and Wadsworth, 1940; Hedges, 1967; Jarvis, 1989) and computer simulation (Dahms, 1992) to collaborative empirical estimates. The possibilities have been reviewed in detail in Section 7.3.

(c) Control charts

s-Charts to control variability. Different symbols for different media should prove advantageous if plotted on the same graph.

8.3.4. Intensified quality control tests

Laboratories differ. Some university and research laboratories do no routine analyses at all and lack therefore easy opportunities to collect enough control data in their daily work. They should make it a routine always to use parallel plates in their analyses. They should also take every opportunity to make note of colony counts on different volumes. They, too, should make it a practice to count the same plate twice occasionally. In this way they would gather data for calculating AQC indices and plotting them on control charts as described under Section 8.2.3.

If this is not enough, then experiments specially designed for intensified testing of analytical performance should be employed. They may be carried out sporadically to control chosen methods and personnel. Also, routine laboratories may resort to them when daily control measures give alarming signals. They may also be useful for training.

The approach to intensified testing is to carry out a design where a split homogeneous suspension is diluted in narrow steps (1:2) and plated in parallel (Weiss et al., 1991). The International Dairy Federation has recently published a standard describing such a test (International Standard IDF 169, 1994. Quality control in the microbiological laboratory, analyst performance assessment for colony count). The test assesses analyst and method performance as well as the technique of dilution.

8.4. Third-line checks

Third-line checks are the responsibility of laboratory management, and are supervised by the quality assurance officer. To assure comparability of results between laboratories, two approaches can basically be used: external quality assessment (EQA) schemes and certified reference materials (CRMs). In both instances, the role of the individual laboratory is to follow exactly the protocol provided by the organizing body and make the calculations as instructed. The analysis and interpretation of the results rests with the central body in the case of EQA, and with the individual laboratories when CRMs are used.

In external quality assessment schemes (proficiency testing), one or more samples from a well-mixed and stable batch are examined by different laboratories and the results are interpreted retrospectively. The central organization collects and analyses the data and informs the individual laboratories of their performance compared with the other participants. This is a flexible approach and it is important that the preliminary results, final report and assessment are returned to the individual laboratories as quickly as possible after the examination date so that the reasons for deviating results can be investigated promptly. Detailed protocols of all operational features are necessary in order to identify these, and to take appropriate action.

The data of laboratory performance tests are in principle suitable for calculating an estimate of the between-laboratory variance of the method, provided that the group of laboratories is a suitable one and that the material variance can be accounted for.

Certified reference materials are stable and homogeneous reference samples, one or more of the properties having been defined by an inter-laboratory study. Ideally, such a study should involve different methods using different measurement principles and traceable to primary standards. Such an approach is not possible in microbiology because all results are method-defined (see Section 7.1.3). Therefore, an alternative approach is chosen where all laboratories use the same method, following a strict protocol and working under carefully controlled conditions. The certified value is then valid only for the applied method. Results obtained with other methods can be compared with the certified value to evaluate the relative recovery of the different methods for that particular type of sample. It should be the responsibility of the central laboratory to provide the detailed protocol for carrying out the tests. It should also provide either the statistical service or the necessary estimates of errors to enable the laboratory to do the necessary calculations to construct control charts.

The CRMs can be analysed when desired and, if necessary, repeatedly to find reasons for deviating results, so that action can be taken. At present, only a limited number of CRMs is available, however. An intermediate form would be that a central laboratory analyses a sample, and the test result is given as a substitute certified value.

References

Dahms, S., 1992. Simulation als Mittel der Modellkritik: über den Versuch ein Problem mikrobiologischen Arbeitens statistisch zu lösen, Diss., Univ. Bielefeld. ISBN 3-89473-479-5.

Dufour, A., 1980. Standard practice for establishing performance characteristics for colony counting methods in bacteriology. ASTM Standard, Section 11, Water and environment, Vol. 11.02.

Eisenhart, C. and Wilson, P.W., 1943. Statistical methods and control in bacteriology. Bacteriol. Revs., 7: 57–137.

Fowler, J.L., Clark, Jr., W.S., Foster, F.F. and Hopkins A., 1978. Analysis variation in doing the standard plate count, as described in: Standard methods for the examination of dairy products. J. Food. Prot., 41: 4–7.

Gameson, A.L.H., 1983. Investigations of Sewage Discharges to Some British Coastal Waters, Chapter 3, Bacteriological Enumeration Procedures, Part 2, Technical Report TR 193, Water Research Centre.

Hedges, A.J., 1967. On the dilution errors involved in estimating bacterial numbers by the plating method. Biometrics, 23: 158–159.

Heisterkamp, S.H., Hoekstra, J.A., van Strijp-Lockefeer, N.G.W.M., Havelaar, A.H., Mooijman, K.A., in 't Veld, P.H. and Notermans, S.H.W., 1992. Statistical analysis of certification trials for microbiological reference materials. Commission of European Communities. Community Bureau of Reference, Brussels.

Hurley, M.A. and Roscoe M.E., 1983. Automated statistical analysis of microbial enumeration by dilution series. J. Appl. Bacteriol., 55: 159–164.

Jarvis, B. 1989. Statistical Aspects of the Microbiological Analysis of Foods. Progress in Industrial Microbiology, Vol. 21. Elsevier, Amsterdam.

Jennison, M.W. and Wadsworth, G.P., 1940. Evaluation of the errors involved in estimating bacterial numbers by the plating method. J. Bacteriol., 39: 389–397.

Klee, A.J., 1993. A computer program for the determination of most probable number and its confidence limits. J. Microbiol. Meth. 18: 91–98.

de Man, J.C., 1975. The probability of the most probable numbers. Eur. J. Appl. Microbiol., 1: 67–78.

de Man, J.C., 1983. MPN tables corrected. Eur. J. Microbiol. Biotechnol., 17: 301–305.

Mooijman, K.A., in 't Veld, P.H., Hoekstra, J.A., Heisterkamp, S.H., Havelaar, A.H., Notermans, S.H.W., Roberts, D., Griepink, B. and Maier, E., 1992. Development of microbiological reference materials, Report EUR 14375 EN, Commission of the European Communities, Luxembourg.

Nordic Committee on food analysis (NMKL), 1989. Handbook for microbiological laboratories. Introduction to internal quality control of analytical work, Report No 5, ISSN 0281-5303.

Schijven, J.F., Havelaar, A.H and Bahar, M., 1994. A simple and widely applicable method for preparing homogeneous and stable quality control samples in water microbiology. Appl. Environ. Microbiol., 60: 4160–4162.

Sen, P.K., 1964. Tests for the validity of the fundamental assumption in dilution

(-direct) assays. Biometrics, 20: 770–784.

Stearman, R.L., 1955. Statistical concepts in microbiology. Bacteriol. Rev., 19: 160–215.

Taylor, J., 1962. The estimation of numbers of bacteria by ten-fold dilution series. J. Appl. Bacteriol., 25: 54–61.

Tillett, H.E. and Coleman, R., 1985. Estimated numbers of bacteria in samples from non-homogeneous bodies of water. J. Appl. Bacteriol., 59: pp. 381–388.

Tillett, H.E., Lightfoot N.F. and Eaton, S., 1993. External quality assessment in water microbiology: statistical analysis of performance. J. Appl. Bacteriol., 74: 497–502.

Tillet, H.E., 1995. Comment on: The most probable number estimates and the usefulness of confidence intervals. Water Res., 29: 1213–1214.

Tillett, H.E. and Lightfoot, N.F., 1995. Quality control in environmental microbiology. Compared with chemistry, what is homogeneous and what is random? Water Sci. Technol., 31: 471–477.

Van Dommeler J.A., 1995. Statistical aspects of the use of microbiological (certified) reference materials. National Institute of Public Health and Environmental Protection, Bilthoven. Report No 281009009, May 1995.

Weiss, H., Niemelä, S. and Arndt, G., 1991. Sicherung der Präzision standardisierter mikrobiologischer Untersuchungsverfahren. Biometrie und Informatik in Medizin und Biologie, 22(3): 116–135, ISSN 09 34-9235.

Chapter 9

Handling and reporting results

9.1. Introduction

This chapter deals with the recording of final results, reporting information from these results, making decisions which may be necessary in the light of these results and subsequent actions which may be needed. A summary is given in Fig. 9.1 (p. 202).

When all tests on a sample have been completed, the final result must be recorded, interpreted and used appropriately. Table 9.1 summarizes different types of samples and then gives a check list of corresponding actions which should be considered.

9.1.2. Recording results

This first action from the list in part B of Table 9.1, recording and checking the final result, applies to all food and water samples handled in a microbiology laboratory. The methods of recording results may or may not be computerized, but the methods should be written down in the laboratory procedures. The procedures should be available to all members of staff at all times.

All results must be recorded in a predetermined place or places in the laboratory records and in a predetermined format. It must be possible to link the result back to information about all stages of receiving and processing the sample. The final result should be checked for clerical or other errors and then assessed for further

Table 9.1

Types of sample and reference list of actions

A.	*Classification of samples*	
	Type of sample	Action check list for cross reference with B
	Internal QC (e.g. first- and second-line checks)	1,2,3
	External QA (e.g. third-line checks)	1,2,3,4,8
	Research	1,2,8
	Routine surveillance	1,2,4,5,6,7,8
	Re-sampling	1,2,4,6,7,8
	Investigational	1,2,4,5,6,7,8

B. *Procedures and actions that may be required*
Action:

1. Record and check the final result in laboratory files;

2. Add interpretation e.g.:   (i) assessment of laboratory findings in the context of accuracy of sampling methods and laboratory procedures;

    (ii) relevance of organisms found;

3. Respond within the laboratory;

4. Report to customer;

5. Request other samples;

6. Report to other agencies;

7. Initiate public health actions;

8. Link with other results and interpret trends, etc.

actions. A designated member of staff should sign this final report form when satisfied that it is correct and that the laboratory had been operating in a satisfactory way, as defined by its quality assurance programme.

9.1.3. Further handling of results

Any subsequent stages of handling of results depend on the type of sample. These fall roughly into two groups, which will be discussed separately. These groups are, firstly, samples generated within the laboratory or requested by the laboratory for checking or research, and, secondly, samples submitted to the laboratory by 'customers'.

9.2. Samples organized or requested internally

These are all the samples used for monitoring performance within the laboratory, and samples from internal or collaborative research programmes. They include the following categories.

9.2.1. Internal quality control samples (first-, second- or third-line checks)

The result should be checked to see whether it is as expected. For example, it should be negative if the sample was a blank check; it should confirm as the correct organism if the sample was a positive control. If a counting procedure was involved, then the result should be compared with the expected range of acceptable results, although it must be remembered that chance variability in the numbers of organisms present in a sample can occasionally produce an out-of-range count which was in fact correct. Such an occasional 'failure' is a statistical inevitability in microbiology, unlike chemistry. The proportion of out-of-range counts which can be blamed on chance will be predetermined by the quality control scheme. More frequent failures will require explanation.

A record (e.g. log book) of QC checks and the results of any subsequent investigations should be kept.

As was described in Chapter 8 (analytical quality control), there are various ways of allowing for the variability in organism distribution between samples and building this into the QC analyses. Sequential plots of results are helpful. Statistical assessment should be applied with attention to the cautions described in earlier chapters. Samples which are certified reference materials will have accompanying instructions for assessing results. Other samples used in the laboratory for quantitative QC will have been processed to check specific procedures, such as replicate samples to test for consistent processing, samples at different dilutions to test for accuracy in dilution or samples processed by different operators to check comparability. Examples of simple and more complex checking programmes are given in Chapter 8. If any QC programme produces more failures than would be expected by chance, then the person in charge of the laboratory must be informed and a list of investigations (of equipment, media, procedures, recording, etc.) performed. When a specific problem is detected, such as a faulty incubator, it will mean that all results achieved by the laboratory while that fault was present become suspect. Repeat samples may have to be requested once the problem has been rectified.

There will be no known correct result for a quantitative QC sample, although a target average may be known. This is an important difference from chemistry QC and means that the observed variation in counts cannot be attributed to measurement errors in the laboratory (Tillett and Lightfoot, 1995). The difficulty in distinguishing any unacceptably large measurement error from the inevitable, natural sampling error is the main reason why QC methods used by chemists (such as control charts) cannot be adopted without making modifications to the statistics or the interpretation. The aim will be to detect poor results when they matter, but not to conclude that things were wrong when in fact the laboratory was 'in control'.

This is a growing area in microbiology QC and laboratories are encouraged to try the methods referred to in earlier chapters and to share their experience (e.g. by publications). This should help the development and refinement of QC in the future.

9.2.2.   *External quality control samples*

A programme of internal QC/QA tests in a laboratory can check for consistent performance. Checks for accuracy of some procedures are possible, for example, with reference materials, but the most objective way of checking performance is to take part in an external quality assessment (EQA) scheme. Such schemes have been developed in some countries and are described in Chapter 8.

EQA samples should be processed as part of the laboratory routine and the result reported as requested by the organizers. A detailed log of participation and results, together with assessment reports received from the organizers, should be accumulated and the laboratory's performance monitored. If poor performance is detected either by the organizers or by the laboratory staff, then a reason must be sought. (The poor performance should be of microbiological and not just statistical significance, for example a large number of results gives the statistical power to detect as significant very small average differences between laboratories). Most schemes allow confidential discussion with the organizers so that experience can be shared in detecting causes and taking remedial action. These steps should also be recorded in the log. In some countries there are independent inspectors, such as accreditors, who have to see these logs.

9.2.3. *Samples examined as part of a research programme*

These should be recorded in the format set down in the study protocol. The required numbers of results are accumulated, then submitted for analysis.

9.3. Samples submitted to the laboratory

In many laboratories, the majority of samples will be submitted by 'customers' from outside the laboratory. The value of the information from such results depends fundamentally on the sampling plan, collection procedures and transport of samples. These factors should, if possible, be discussed with the customer.

All samples should be identifiable on arrival at the laboratory, preferably with a request form specifically designed by the laboratory. (The request and report form can be separate or can be integrated into one document — see the examples in Section 9.4). The laboratory should refuse to accept any samples of completely unknown origin, for reasons of safety (e.g. they may contain pathogens) as well as impracticability. Samples arriving with an incomplete history or inadequate labelling and paperwork, should be queried and processed only after satisfactory clarification, unless there is a contract obligation to process the sample regardless of the information supplied.

Examples of report forms are given in Section 9.4. Some have provision for interpretational comment and require two signatures.

9.3.1.   *Routine surveillance samples*

(a) Examined for the presence of pathogens

A negative result should be reported as 'none found in the sample portion examined'. This wording is appropriate because it is usually practical to examine only a small part of the food or water source and, because of variability in organism distribution, the sample taken may be negative when some parts of the source are in fact contaminated. However, for statutory compliance samples there may be a prescribed format for wording the report which has to be used.

Reporting a nil finding in terms of 'less than' a limit of detection does not have the relevance in microbiology that it has in chemistry, as is discussed in the next section on routine samples examined for indicator organism counts.

If an organism is detected which may be of public health significance, then the laboratory should follow pre-planned actions which balance risks to health against the cost of false alarms. These might be:

(i)   to alert the customer to the potential problem and to keep them informed of the results of confirmatory tests on the organism;

(ii)   to consult a medical microbiologist;

(iii)   to ensure that anyone who should be informed is informed. If an organism of potential public health hazard is confirmed from a product prepared for direct consumption then someone will have to decide, with the cooperation of the customer, on whether warnings should be issued and whether surveillance of possible related illnesses should be set up.

The reports to the customer, preliminary and/or final, will describe the organisms found, give counts if applicable and the microbiological interpretation (seeking medical microbiological interpretation from outside if none is available within the laboratory). This interpretation may be omitted if it is not part of the contract with the customer, but a responsible microbiology laboratory will ensure that someone is taking action by responding to public health issues relating to a pathogen which is a potential hazard (found in pre-cooked food or pre-treated water) or a definite hazard (found in the product offered for consumption). Responsibilities may include repeat sampling, notifying other agencies such as the product producer (e.g. food manufacturer, water company), notifying health and environmental officials who will consider actions in the interest of consumers. In some countries, failure to be party to responsible actions may have legal implications. The contract between the customer and the laboratory usually includes a guarantee of confidentiality. The possible conflict of interest which may result when a health hazard is detected should be discussed with customers in advance of any problem.

(b) Examined for total or indicator organism counts (including presence/absence, which is regarded as counts of 0 and 1+)

The result will be reported to the customer as a nil finding or as a count. Consideration should be given to qualifying the count with a comment on accuracy and an assessment of the significance of the finding.

A nil finding should usually be reported as a zero count or as 'none found in the sample portion examined'. If the amount examined is less than the unit of the standard format of report, for example if 10 ml of water was examined and the organism count is to be expressed as the number per 100 ml, then it has sometimes been the custom that the reported result should reflect a 'limit of detection', in this case 10. That assumes that organisms are evenly distributed through the water and that you would expect 1 organism in 10 ml if there are 10 in 100 ml. Random variation, at the very least, will be observed in the distribution of organisms, which means that the 95% confidence intervals for a 100 ml portion when 0 was found in 10 ml is 0–36. Random variation is the smallest variability that can be expected. It may be achieved with very well-mixed samples. Therefore, a zero count that is converted into a '<' statement is not recommended, because it implies a greater knowledge of what was present in the product than is actually known. However, as discussed in Section 9.3.1(a), in some

countries there may be situations where this recommendation cannot be followed because there is a prescribed format for reporting on
statutory sampling.

A positive finding should either be reported as 'organisms present', or as a count for the volume or weight examined or as a count
rescaled to a conventional unit of weight or volume.

It can be illuminating to consider the accuracy of the count from a
sampling and statistical point of view. Publications show that this is
a complex area. There are two sources of potential uncertainty —
variation in organism density at the source from which the sample was
taken, and imprecision introduced by laboratory techniques such as
sub-sampling and diluting, when random variation can have surpris-
ingly large effects on the results (Anon., 1994). However, if the laboratory techniques are being properly practised and are being shown as
satisfactory by QC/QA checks, then the reported count will be the best
the laboratory can provide and should be an unbiased estimate of what
was in the sample. Thus, in setting up the QC/QA tests described in
Chapter 8, great use is made of the theory and observation of imprecision due to laboratory techniques.However, when reporting a result
to a customer, whether the laboratory is using a counting or an estimation method, it is probably better not to use this theory to add any
statement about inaccuracy introduced by laboratory procedures, for
example by giving a 95% confidence interval (ci), because these may
be misunderstood. The customer may think the statement applies to
the source and not to the sample. Only multiple samples can give sufficient information to allow confidence intervals to be estimated for
the microbial density at the source, where the variation in microbial
content is often very large (Tillett, 1993). The planning of routine
sampling schemes is outside the scope of this book, but, in principle,
the greater the variation at the source then the greater the number
of samples needed to give reliable information.

If there was uncertainty about the microbial density and more
than one dilution was used for a counting method, then there may be
occasions when two or more counts can be made from different dilutions. In that case the result should be given either as:

(i)   an aggregated count divided by the total of the volumes in-
        volved. For example, if 100 ml of the sample yielded a count of

148 and a 10-fold dilution (i.e. 10 ml of the sample) yielded a count of 49 then the result would be:

$$\frac{148 + 49}{100 + 10} = 1.79 \text{ per ml}$$

which would probably be reported as 179 per 100 ml. (Further examples are given in the International Dairy Federation Standard 145.); or

(ii) the count chosen which is within or closest to, a prescribed range of acceptable accuracy. For example, a membrane colony count of between 20 and 80 organisms is recommended for counting indicator organisms from water (Anon., 1994).

Occasionally a count cannot be achieved, for example, when there was insufficient diluting of the sample resulting in overcrowding of colonies or growth in all portions in a dilution series. An unsatisfactory result can also occur if a selective medium fails and there is overgrowth of irrelevant organisms which obscure the relevant colonies. When these deficient results occur, the report to the customer should reflect what has been observed, where the deficiency lies, what can be inferred, if anything, and, if appropriate, a repeat sample should be requested, bearing in mind that it will be impossible to obtain an identical one. For example, with growth in all portions of a dilution series then it is correct to report the count as 'probably greater than an estimated number'.

Thus, a simple statement of 'none found' or a count per unit (e.g. g or 100 ml) is recommended as the reported result from a sample which has been processed satisfactorily.

As described in the previous section, when examining for pathogens, the laboratory may or may not be contracted to add interpretation to the result. However, routine surveillance is done to check whether the counts are satisfactory or not. At the very least, the laboratory should ensure that someone is taking responsible action if the results indicate a potential health hazard.

9.3.2.   Samples for statutory monitoring

These samples may need to be examined for pathogens, and for total or indicator organism counts. All the comments for Sections (a) and (b) in the previous section apply. In addition, it will be necessary to ensure that the result is reported in accordance with any relevant national or international legislation.

If the sample was part of a statutory monitoring programme, then the report should be highlighted as such in the laboratory's records. It will be reported to the appropriate authorities, but may first have to be categorized as 'pass' or 'fail'.

Some regulations are worded such that most, but not all, samples must pass the standard. Such a regulation may be described as a percentage compliance over a stated period of time. For this type of monitoring, the results for the time period will be accumulated and the appropriate percentile of the results must be calculated or estimated. This percentile count will then be reported as below or above the standard. Percentage compliance is not as straightforward as it sounds because there are different approaches used to estimate percentiles. When the total number of samples is large, this is less of a problem, but if they are very much fewer than 100, there may be no result close to the required percentile and a method of estimating should be agreed with the regulators. This may be a simple statement of the number of allowed 'failures' for given numbers of samples processed, or it may be a statistical estimation of the equivalent percentile.

For example, if 12 samples are examined and the 95 percentile is required, the first step is to rank the counts in ascending order. The 11th ranked count is the 91.67 percentile and the 95th will be estimated to be between the 11th and 12th ranked counts. The counts can be plotted in rank order and if they appear approximately linear then interpolation between the top two observations ($x11$ and $x12$) will estimate the 95 percentile, using:

$$x11 + \frac{(95 - 91.67)}{(100 - 91.67)} \cdot (x12 - x11)$$

With microbiological counts, this approximate linearity may require the use of a logarithmic scale. The linear interpolation is then performed on this scale and the result converted back to the original scale. The following example illustrates this with 12 total coliform counts from samples taken from a bathing beach in one season. The results were: 6700, 994, 1062, 424, 2900, 140, 30, 190, 230, 150, 100, 1500. If these are plotted in ascending order on a logarithmic scale, then they lie approximately on a straight line.

The logarithms of the two highest values are 3.4624 and 3.8261 and so the 95 percentile is estimated to be the antilogarithm of:

$$3.4624 + \frac{(95 - 91.67)}{(100 - 91.67)} \cdot (3.8261 - 3.4624) = 3.608$$

giving an estimated 95th percentile of 4050 coliforms per 100 ml, rounded to the nearest 10.

There may be valuable information which can be obtained by monitoring these results in addition to the information reported for statutory purposes, because the 'pass/fail' comparison is a very simplistic interpretation in some situations. Laboratories or customers may therefore wish to keep serial records of all this information to study patterns. This might include graphs illustrating average and variation, plotted over time.

9.3.3. Serial routine surveillance samples

Routine testing of food or water is often organized to monitor indicator or total organism counts over time. Thus, repeated samples from the same food or water source, at the same stage of processing (e.g. a raw water source at the point of abstraction or a sampling point where water leaves the treatment works) need to be analysed for trends. If it is the responsibility of the laboratory to assemble and interpret the results from this source then the storage of data in the laboratory must enable the linking of serial results to be done easily, and some statistical methods for detecting patterns and trend should be planned.

Graphical plots of organism counts against time are essential. Logarithmic or square root scales may be needed to cope with high average counts, but if the usual count is low or zero then an arithmetic scale gives a more informative picture. From these graphs it

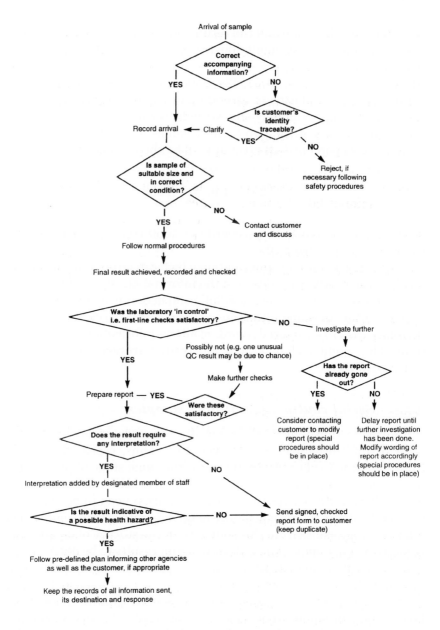

Fig. 9.1. Flow chart for reporting customers' samples (use in conjunction with Fig. 1.1 and Fig. 4.1).

may be obvious what the background variability is, whether there is any seasonal pattern, whether there are upward or downward trends. Further assessment using statistics may be needed to confirm the obvious, or to detect more subtle changes. Because microbiological counts tend to be highly variable, it may take quite a long series of routine results before the background variation can be measured at all accurately.

Averages, usually medians, from adjacent time periods, or moving averages rolled forward over time are the simplest approach to trend analysis. More complex modelling, such as the statistical process control techniques used by some water companies can be applied. They are often very sensitive to change and underestimate the inevitable high variability of counts, but they are a useful tool for screening for possible changes. Whatever analysis is applied, there should be clear objectives laid down at the planning stage about what needs to be detected and what the reaction should be.

9.3.4. Non-routine samples (for pathogens, total or indicator organisms)

(a) Follow-up

These are taken in response to unsatisfactory routine samples. The results should be reported in a way which links them clearly with the results from the first sample and, if appropriate, a statement should be added to say whether the counts are now satisfactory and whether or not more sampling is advisable.

(b) Investigational

Samples that are taken in response to a suspected problem will include those where there is illness in the consumers, where there has been a known breakdown in the handling or treating of the product or a problem in distribution, such as repair work to water supply pipes. These samples will need to be processed and reported rapidly and there should be a mechanism within the laboratory procedures for sending out preliminary as well as final reports. If the results provide evidence that there is a hazard, the laboratory should ensure that responsible action is taken.

(c) Other/and ad hoc

There may be samples other than those already discussed. The same principals apply. The laboratory should supply accurate and informative reports and any results suggestive of a public health hazard should be noted and discussed.

9.4. Examples of request forms and report forms

Examples of request forms and report forms are reproduced opposite and on the following pages.

References

Anon., 1994. The Microbiology of Water (1994). Part 1 — Drinking Water. Report on Public Health and Medical Subjects, No. 71. HMSO, London.

Tillett, H.E., 1993. Potential inaccuracy of microbiological counts from routine water samples. Water Sci. Technol., 27: 15–18.

Tillett, H.E. and Lightfoot, N.F., 1995. Quality control in environmental microbiology compared with chemistry: what is homogenous and what is random? Water Sci. Technol., 31: 471–477.

PUBLIC HEALTH LABORATORY SERVICE

PUBLIC HEALTH LABORATORY
INSTITUTE OF PATHOLOGY, GENERAL HOSPITAL,
WESTGATE ROAD, NEWCASTLE-UPON-TYNE NE4 6BE
Telephone : (091) 273 8811 Ext. 22806
Fax No: (091) 226 0365

BACTERIOLOGICAL EXAMINATION OF FOOD SAMPLES

NAMAS
TESTING
No. 1617

FOR LABORATORY USE
CATEGORY

Routine control
Tick as appropriate

Food poisoning
(see overleaf)

Sender _____

Sampled by _____

Date of collection _____

Address _____

Date received _____

Time received _____

Received by _____

Additional information/requests

Senders Reference Number	Laboratory Reference Number	Product Date: Purchased Produced Use by	Origin	Aerobic Plate Count cfu/g	Staph. aureus/g	Bacillus cereus/g	E.coli/g	Salmonella species/25g	Campylobacter species/25g

Additional results / comments :

Telephoned _____ Date reported _____ Microbiologist _____

Methods ref : "Newcastle Public Health Laboratory Food Methods" Current Issue.
Food Form Issue 4 - 24.5.96.

Prof.R.Freeman, Dr.N.F.Lightfoot, Dr.A.A.Codd, Dr.A.Galloway.

Page 1 of 1

PUBLIC HEALTH LABORATORY SERVICE

FOR LABORATORY USE

CATEGORY

PUBLIC HEALTH LABORATORY
INSTITUTE OF PATHOLOGY, GENERAL HOSPITAL,
WESTGATE ROAD, NEWCASTLE-UPON-TYNE NE4 6BE
Telephone : (091) 273 8811 Ext. 22806
Fax No: (091) 226 0365

NAMAS
TESTING
No. 1617

BACTERIOLOGICAL EXAMINATION OF WATER SAMPLES

Sender _____

Sampled by _____

Address _____

Date of collection _____

Date received _____

Time received _____

Received by _____

Senders Reference Number	Laboratory Reference Number	SAMPLE DESCRIPTION	Total Coliforms	Thermo-tolerant Coliforms	E. coli	Faecal Streps	Pseudo. aeruginosa	Sulphite reducing Clostridia	22.C/72hrs	37.C/24hrs
					Membrane Filtration				Colony Count	
		Sample taken from _____ Time of collection** _____ Class (Private water) Reason for testing Routine ☐ Follow up ☐ Complaint ☐ * Chlorinated ☐ Filtered ☐ Untreated ☐ * Additional tests _____	per 100ml	per 100ml	per 100ml	per 100ml	per 100ml		per ml	per ml
		Sample taken from _____ Time of collection** _____ Class (Private water) Reason for testing Routine ☐ Follow up ☐ Complaint ☐ * Chlorinated ☐ Filtered ☐ Untreated ☐ * Additional tests _____	per 100ml	per 100ml	per 100ml	per 100ml	per 100ml		per ml	per ml
		Sample taken from _____ Time of collection** _____ Class (Private water) Reason for testing Routine ☐ Follow up ☐ Complaint ☐ * Chlorinated ☐ Filtered ☐ Untreated ☐ * Additional tests _____	per 100ml	per 100ml	per 100ml	per 100ml	per 100ml		per ml	per ml

* Tick as appropriate

Telephoned _____   Date reported _____   Microbiologist _____

Methods ref: "The microbiology of water 1994 part 1 Drinking water, Report on Public Health and Medical Subjects No 71.   Prof.R.Freeman, Dr.N.F.Lightfoot, Dr.A.A.Codd, Dr.A.Galloway.

N.B. Analysis of water samples must commence within 6 hours of collection.

Water Form Issue 4 - 24.5.96.   Page 1 o

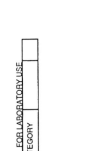

PHLS

NAMAS
TESTING
No. 1617

PUBLIC HEALTH LABORATORY SERVICE

PUBLIC HEALTH LABORATORY
INSTITUTE OF PATHOLOGY, GENERAL HOSPITAL,
WESTGATE ROAD, NEWCASTLE-UPON-TYNE NE4 6BE
Telephone : (091) 273 8811 Ext. 22806
Fax No: (091) 226 0365

BACTERIOLOGICAL EXAMINATION OF MILK SAMPLES

Dairy ....

Sender ....   Date received ....

Sampled by ....   Time received ....

Date of collection ....   Received by ....

Address ....

Senders reference number	Laboratory reference number	Pasteurised Milk Type (circle as applicable)	Package Size	Temperature on arrival	Coliforms / ml 30.C 24hrs		Plate count / ml 21.C 25hrs (following pre-incubation 6.C 120hrs)		Phosphatase test
					Count	Standard Exceeded Y/N	Pass/Fail	Count	Standard Exceeded Y/N	Pass/Fail	37.C 2hrs Pass/Fail
		WHOLE SEMI-SKIMMED SKIMMED								

Previous counts - record when applicable
1 =
2 =
3 =
4 =

Previous counts - record when applicable
1 =
2 =
3 =
4 =

Key

P = Passed prescribed test

F = Failed prescribed test

V = Void test (>4.C or <0.C on arrival)

N/A = Not applicable

Y = Yes

N = No

Standard for Coliforms:-
n=5, C=1, m=0, M=5

Standard for Plate Count:-
n=5, C=1, m=5x10^4, M=5x10^5

Where n = The number of sample units comprising the sample i.e. 5 submissions to the laboratory.
c = The number of sample units where the bacterial count may be between m and M, if the other bacterial counts are m or less. The test fails if any sample count exceeds M.
m = The threshold value for the number of bacteria, result satisfactory if not exceeded.
M = The maximum count value, result unsatisfactory if exceeded in any sample unit.

Telephoned ....   Date reported ....   **Microbiologist** ....

Methods ref: "Newcastle Public Health Laboratory Milk Methods" Current Issue.   Prof.R.Freeman, Dr.N.F.Lightfoot, Dr.A.A.Codd, Dr.A.Galloway.
Milk Form Issue 4 - 24.5.96.

Page 1 of 1

FOR LABORATORY USE
CATEGORY

PHLS

NAMAS
TESTING
No. 1617

PUBLIC HEALTH LABORATORY SERVICE

PUBLIC HEALTH LABORATORY
INSTITUTE OF PATHOLOGY, GENERAL HOSPITAL,
WESTGATE ROAD, NEWCASTLE-UPON-TYNE   NE4 6BE
Telephone : (091) 273 8811 Ext. 22806
Fax No: (091) 226 0365

BACTERIOLOGICAL EXAMINATION OF MOLLUSCS

FOR LABORATORY USE

CATEGORY

Sender

Address

Sampled by

Date received

Date of collection

Time received

Received by

Senders Reference Number	Laboratory Reference Number	Product	Origin	Results	
				Faecal Coliforms / 100g	E. coli / 100g

Telephoned   Date reported   Microbiologist
Prof.R.Freeman, Dr.N.F.Lightfoot, Dr.A.A.Codd, Dr.A.Galloway.

Methods ref: "Newcastle Public Health Laboratory Food Methods" Current Issue.
Mollusc Form Issue 4 - 24.5.96.

Page 1 of 1

PUBLIC HEALTH LABORATORY SERVICE

PUBLIC HEALTH LABORATORY
INSTITUTE OF PATHOLOGY, GENERAL HOSPITAL,
WESTGATE ROAD, NEWCASTLE-UPON-TYNE NE4 6BE
Telephone : (091) 273 8811 Ext. 22806
Fax No: (091) 226 0365

NAMAS
TESTING
No. 1617

FOR LABORATORY USE

CATEGORY

EXAMINATION OF WATER SAMPLES FOR LEGIONELLA

Sender _____

Address _____

Sampled by _____

Date of collection _____

Date received _____

Time received _____

Received by _____

Senders Reference Number	Laboratory Reference Number	NATURE OF SAMPLE	RESULTS AND COMMENTS
		Sample taken from _____ Total Capacity of System _____ Temperature _____ Untreated ☐ Filtered ☐ Tick as appropriate _____ Biocide _____ Biocide Conc. _____	
		Sample taken from _____ Total Capacity of System _____ Temperature _____ Untreated ☐ Filtered ☐ Tick as appropriate _____ Biocide _____ Biocide Conc. _____	
		Sample taken from _____ Total Capacity of System _____ Temperature _____ Untreated ☐ Filtered ☐ Tick as appropriate _____ Biocide _____ Biocide Conc. _____	

When Legionella organisms have NOT been isolated from the samples submitted, it should be remembered that this only reflects the situation on the day of the test
Careful maintenance programmes are required to prevent Legionella colonisation.
We reserve the right to inform the appropriate authorities of any positive isolates that have Public Health importance.

Telephoned _____ Date reported _____ Microbiologist _____

Methods ref : "Newcastle Public Health Laboratory Legionella Methods" Current Issue   Prof.R.Freeman, Dr.N.F.Lightfoot, Dr.A.A.Codd, Dr.A.Galloway.
Legionella Form Issue 4 - 24.5.96.   Page 1 of :

PUBLIC HEALTH LABORATORY SERVICE

PUBLIC HEALTH LABORATORY
INSTITUTE OF PATHOLOGY,
NEWCASTLE GENERAL HOSPITAL,
WESTGATE ROAD,
NEWCASTLE UPON TYNE NE4 6BE
Telephone: (091) 273 8811
Fax No: (091) 226 0365

MICROBIOLOGICAL EXAMINATION OF FOOD

NAMAS TESTING No.1617

Sampling Authority _____

Address _____

Report (send to) _____

Contact telephone _____ Fax _____

Print ALL details

Sample Ref No _____

Sample received from _____

Sample collected by _____ Status _____ (ref _____

Sample received by _____

LAB NO _____

Sample received on _____ Temperature on receipt _____ °C

Purpose of investigation: Formal ☐ Outbreak ☐ Other ☐

Storage conditions since receipt by laboratory _____ (_____ °C)

Sample: Suspect item ☐ Item from same batch ☐ Type _____

Laboratory investigations commenced at _____ am/pm

(In suspected food poisoning cases also complete details overleaf)

Date of sampling _____ Time _____ Approximate Weight _____ g Batch/Lot no _____

Appearance _____

Microscopy _____

Process code _____ Use by date _____

Plate count, Aerobic/g _____ pH _____

Place of sampling _____ (Post code _____

     at _____ °C for _____ hours

Anaerobic/g _____ at _____ °C for _____ hours

     at _____ °C for _____ hours

Manufacturer ☐ Importer ☐ Wholesaler ☐ Retailer ☐ Other _____ (Code _____

Coliforms/g _____ E.coli/g _____

Salmonella spp/25g _____ Ident _____

Name of Producer/Wholesaler (for retailed sliced meats, etc.) _____

Listeria spp/25g _____

Sample collected from: Shelf ☐ Cabinet ☐ °C Fridge ☐ Freezer ☐ Other ☐

Staph.aureus/g _____ C.perfringens/g _____

Storage condition at place of sampling: Temp _____ °C Humidity _____ % Sanitation _____

Campylobacter spp /25g _____ Ident _____

Enterococci /g _____ Bacillus cereus/g _____

Condition of packaging: Clean whole & intact ☐ Dirty ☐ Damaged ☐ Leaking ☐ Opened ☐

Vibrio parahaemolyticus/25g _____ Yersinia spp/25g _____

Cooking process _____ Date of cooking _____

Yeasts/g _____ Moulds/g _____

Country of origin _____

Interpretation _____

Transport condition: Time (hrs) _____ Mode of transport _____ Temperature _____ °C

Method of sampling: Random throughout lot ☐ Random throughout accessible units ☐ Isolated sample ☐

Food Examiner _____ Microbiologist * _____

Storage & transport conditions since sample taken _____ (_____ °C)

Reported _____ Certificate of Examination issued Yes ☐ No ☐

Formal Food Form Issue 4 - 24.5.96.

* Prof R.Freeman, Dr N.F.Lightfoot, Dr A.A.Codd, Dr A.Galloway   Page 1

INSTRUCTIONS CONCERNANT LES ANALYSES D'EAU

1 - LE PRELEVEMENT

a) L'analyse bactériologique : L'échantillon sera recueilli dans un flacon spécial disponible au laboratoire (flacon stérile avec neutralisant). Si le point d'eau se trouve en plein air, se mettre à l'abri du vent et des poussières. S'il s'agit d'une pompe, d'un robinet..., flamber l'orifice et faire couler l'eau durant 5 minutes avant de recueillir l'échantillon. Ouvrir le flacon (en évitant que le contact des doigts ne souille le goulot), remplir d'eau et reboucher soigneusement. Les échantillons conservés en glacière (+ 4°C) doivent être apportés sans délai au laboratoire, au plus tard dans les 8 heures.

b) L'analyse chimique : Recueillir l'eau dans des bouteilles en matière plastique à usage unique, disponibles au laboratoire. Certaines déterminations (oxygène dissous, hydrocarbures, pesticides, etc...) exigent un mode de prélèvement particulier ou des flacons en verre spéciaux. A la moindre hésitation, se mettre en relation avec le laboratoire AVANT de prélever. Les échantillons conservés en glacière (+ 4°C) seront apportés dans les plus brefs délais au laboratoire.

2 - LES ANALYSES

- ANALYSES DE POTABILITE : - Sommaire (A2) : 1 flacon Microbiologie + 1 flacon Chimie
        - Réduite (A1)   : 1 flacon Microbiologie + 1 flacon Chimie
        - Complète   : Consulter le laboratoire

- CONTROLE DE BAIGNADE   : Consulter le laboratoire
- FONTAINES REFRIGEREES
- DISTRIBUTEURS DE BOISSON
- ANALYSE AVANT TRAVAUX DE PLOMBERIE SANITAIRE
- CONTROLES D'ATMOSPHERES ET CLIMATISEURS
- CONTROLES D'EAU POUR HEMODIALYSE
- CONTROLES D'EMBALLAGES, OU PRODUITS INDUSTRIELS
- EFFLUENTS INDUSTRIELS OU URBAINS
- EAUX MINERALES, EAUX CONDITIONNEES
- TESTS DE TOXICITE

et toutes études concernant les eaux, leur qualité, leur traitement, l'environnement

INSTITUT PASTEUR DE LILLE
Service Eaux-Environnement
Laboratoire de référence agréé pour l'analyse des eaux
Ouvert du lundi au vendredi de 8h30 à 12h30 et de 13h30 à 17h45

LILLE :
1, rue Calmette - BP 245 - 59019 LILLE CEDEX
Tél. 03 20 87 77 30 - Fax : 03 20 87 73 83

GRAVELINES :
Route du Colombier - 59820 GRAVELINES
Tél. : 03 28 23 03 61 - Fax : 03 28 23 09 47

DEMANDE D'ANALYSE

A remettre au laboratoire avec le bon de commande
en même temps que les flacons
(PAS DE COURRIER SEPARE - SVP)

DEMANDEUR -
NOM ou RAISON SOCIALE : _____
n°, rue : _____
Complément d'adresse : _____ Commune : _____ Dépt. : ____

PAYEUR -
NOM ou RAISON SOCIALE : _____
n°, rue : _____
Complément d'adresse : _____ Commune : _____ Dépt. : ____
Bon de commande : Oui / Non   Si oui n° : _____
Y-a-t-il une convention avec l'Institut Pasteur de Lille : Oui / Non
Si oui références et date d'effet : _____

DEMANDE -
Prélèvement :
Genre : Eau de réseau, Puits, Forage, Piézomètre, Rivière, Mer, Effluent urbain, Effluent industriel, Eau minérale, Eau conditionnée, Eau stérile, Fontaine réfrigérée, Sol, Fertilisant, Produit industriel, Produit agricole, Aliment,
autre : _____
Commune : _____ Dépt. : ____
n°, rue : _____
Prélevé le : _____ à ___h___ par _____ (km: ____ - Temps : ____)
Nombre de flacons : _____   Flacons Pasteur : Oui / Non
Traitement : Oui / Non   Chlore : Oui / Non   Autres : _____
Remarque (but de l'analyse..) _____

Description des échantillons :

Point de prélèvement :	Analyses demandées :	Résultats de mesures de terrain	n° d'analyse : (Réservé au laboratoire)

Signature du demandeur :

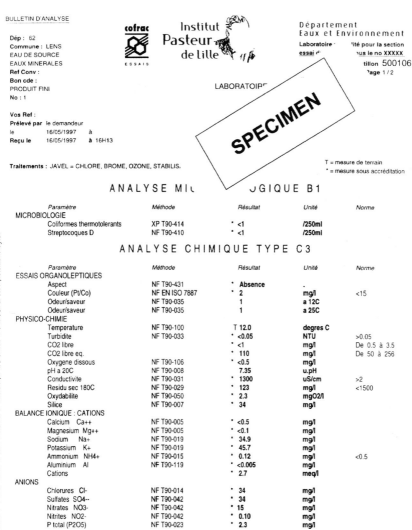

BULLETIN D'ANALYSE

Dép : 62
Commune : LENS
EAU DE SOURCE
EAUX MINERALES
Ref Conv :
Bon cde :
PRODUIT FINI
No : 1

cofrac
ESSAIS

Institut
Pasteur
de Lille

LABORATOIRF

Département
Eaux et Environnement
Laboratoire ' 'ité pour la section
essai r' us le no XXXXX
tillon 500106
'age 1 / 2

Vos Ref :
Prélevé par le demandeur
le 16/05/1997 à
Reçu le 16/05/1997 à 16H13

Traitements : JAVEL = CHLORE, BROME, OZONE, STABILIS.

T = mesure de terrain
* = mesure sous accréditation

ANALYSE MIꞯ ꞯGIQUE B1

Paramètre	Méthode	Résultat	Unité	Norme
MICROBIOLOGIE				
Coliformes thermotolerants	XP T90-414	* <1	/250ml	
Streptocoques D	NF T90-410	* <1	/250ml	

ANALYSE CHIMIQUE TYPE C3

Paramètre	Méthode	Résultat	Unité	Norme
ESSAIS ORGANOLEPTIQUES				
Aspect	NF T90-431	* Absence	.	
Couleur (Pt/Co)	NF EN ISO 7887	* 2	mg/l	<15
Odeur/saveur	NF T90-035	1	a 12C	
Odeur/saveur	NF T90-035	1	a 25C	
PHYSICO-CHIMIE				
Temperature	NF T90-100	T 12.0	degres C	
Turbidite	NF T90-033	* <0.05	NTU	>0.05
CO2 libre		* <1	mg/l	De 0.5 à 3.5
CO2 libre eq.		* 110	mg/l	De 50 à 256
Oxygene dissous	NF T90-106	* <0.5	mg/l	
pH a 20C	NF T90-008	7.35	u.pH	
Conductivite	NF T90-031	* 1300	uS/cm	>2
Residu sec 180C	NF T90-029	* 123	mg/l	<1500
Oxydabilite	NF T90-050	* 2.3	mgO2/l	
Silice	NF T90-007	* 34	mg/l	
BALANCE IONIQUE : CATIONS				
Calcium Ca++	NF T90-005	* <0.5	mg/l	
Magnesium Mg++	NF T90-005	* <0.1	mg/l	
Sodium Na+	NF T90-019	* 34.9	mg/l	
Potassium K+	NF T90-019	* 45.7	mg/l	
Ammonium NH4+	NF T90-015	* 0.12	mg/l	<0.5
Aluminium Al	NF T90-119	* <0.005	mg/l	
Cations		* 2.7	meq/l	
ANIONS				
Chlorures Cl-	NF T90-014	* 34	mg/l	
Sulfates SO4--	NF T90-042	* 34	mg/l	
Nitrates NO3-	NF T90-042	* 15	mg/l	
Nitrites NO2-	NF T90-042	* 0.10	mg/l	
P total (P2O5)	NF T90-023	* 2.3	mg/l	

Fondation reconnue
, d'utilité publique

1, rue du Professeur Calmette
B.P. 245 - 59019 Lille cedex
Tel. 03 20 87 77 30 à 33 · Fax 03 20 87 73 83

Laboratoire de référence agrée
pour l'analyse des eaux

BULLETIN D'ANALYSE

Dép : 62
Commune : LENS
EAU DE SOURCE
EAUX MINERALES
Ref Conv :
Bon cde :
PRODUIT FINI
No : 1

Vos Ref :
Prélevé par le demandeur
le   16/05/1997   à
Reçu le   16/05/1997   à 16H13

cofrac ESSAIS   **Institut Pasteur de Lille**

Département
Eaux et Environnement
Laboratoire accrédité pour la section
essai du COFRAC sous le no XXXXX
Echantillon 500106
Page 2 / 2

LABORATOIRE D'ANALYSES MEDICALES

FRANCE

Traitements : JAVEL = CHLORE, BROME, OZONE, STABILISANT

T = mesure de terrain
* = mesure sous accréditation

CarbonatesCO3--			* <2	mg/l	
Bicarb. HCO3-			* 1500	mg/l	
Fluorures F-	NF T90-004		* <0.05	mg/l	
Anions	NF T90-039		* 26.3	meq/l	
PARAMETRES INDESIRABLES					
Fer total Fe	ISO 11 885		* <0.02	mg/l	
Manganese Mn	ISO 11 885		<0.02	mg/l	
Cuivre Cu	ISO 11 885		* <0.02	mg/l	<1
Zinc Zn	ISO 11 885		* <0.050	mg/l	De 3.8 à 4.9

SPECIMEN

Eau minerale faiblement mineralisee,bicarbonatee.
Presence anormale de Turbidite.
QUALITE CHIMIQUE INSUFFISANTE.

A Lille, le 15/12/1997   Le Chef de Service,

Fondation reconnue
d'utilité publique

1, rue du Professeur Calmette
B.P. 245 - 59019 Lille cedex
Tél. 03 20 87 77 30 à 33 - Fax 03 20 87 73 83

Laboratoire de référence agréé
pour l'analyse des eaux

(texte vertical en marge gauche :) Toute référence à l'Institut Pasteur de Lille est soumise à l'accord exprès, préalable et écrit d'un de ses représentants légaux

BULLETIN D'ANALYSE

Dép : 59
Commune : FLERS EN ESCREBIEUX

cofrac
ESSAIS

Institut
Pasteur
de Lille

Département
Eaux et Environnement
Laboratoire accrédité pour la section
essai du COFRAC sous le no XXXXX

Ref Conv :
Bon cde :

Ste CASTROL

Page 1 / 1

Prélevé par le demandeur

Reçu le   22/08/1997   à 15H00

LE PECQ FRANCE
78230 LE PECQ

Traitements : 1=A1, 2=A2, J=JAVEL = CHLORE, B=BROME, Z=OZONE, U=UV

* = mesure sous accréditation

DIVERS

No échantillon / Point	Méthode : Norme :	Mat. suspension NF T90-105	Carbone COT NF T90-102 De 0 à 1800	DCO (en O2) NF T90-101	DBO5 (en O2) NF T90-103 De 100 à 6500	Indice CH2 NF T90-109 De 0.5 à 15.5	
729316 LIQ 1 Vos ref : Libellé : IMPREX BAIN No 1 Prélevé le 20/08/1997 à 11H20 Traitements : J, B, Z	Résultat : Unité :	* 14 mg/l	* 225 mg/l	* 770 mg/l	* 175 mg/l	* 1.1 mg/l	
729317 LIQ 1 Vos ref : Libellé : IMPREX BAIN No 2 Prélevé le 21/08/1997 à 11H21 Traitements : B, 1	Résultat : Unité :	* 23 mg/l	* 13.0 mg/l	* 5110 mg/l	* 1680 mg/l	* 0.8 mg/l	
729318 LIQ 1 Vos ref : Libellé : IMPREX BAIN No 3 Prélevé le 22/08/1997 à 11H22 Traitements : Z, U	Résultat : Unité :	* 88 mg/l	* 11500 mg/l	* 44100 mg/l	* 4250 mg/l	* 11 mg/l	
729319 LIQ 1 Vos ref : Libellé : IMPREX PRODUIT PUR Prélevé le 23/08/1997 à 11H23 Traitements : 2	Résultat : Unité :		* 550000 mg/l	* 1760000 mg/l			

SPECIMEN

A Lille, le 15/12/1997   Le Chef de Service,

Fondation reconnue
d'utilité publique

1, rue du Professeur Calmette
B.P. 245 - 59019 Lille cedex
Tél. 03 20 87 77 30 à 33 - Fax 03 20 87 73 83

Laboratoire de référence agréé
pour l'analyse des eaux

Toute référence à l'Institut Pasteur de Lille est soumise à l'accord express, préalable et écrit d'un de ses représentants légaux

BULLETIN D'ANALYSE

Dép : 62
Commune : CALAIS
NOUVELLE STATION

  Institut
Pasteur
de Lille

ESSAIS

Ref Conv : AEAPSTA
Bon cde : 96W87

Département
Eaux et Environnement
Laboratoire accrédité pour la section
essai du COFRAC sous le no XXXXX

Page 1 / 1

AGENCE DE BASSIN ARTOIS PICARDIE

Prélevé par Inst. Pasteur (E.R)

Reçu le   28/08/1997   à 17H00

FRANCE
59508 DOUAI

Traitements : 1=A1, 2=A2, J=JAVEL = CHLORE, B=BROME, Z=OZONE, C=CHAINE, S=STABILISANT, P=PHMB   T = mesure de terrain
    * = mesure sous accréditation

STATION D EPURATION

Echantillon no		729694	729695	729696	729697	729698	729699
Vos ref		1	INTERNE	EXTERNE	1	INTERNE	EXTERNE
Remarques		11H00-11H00	11H00-11H00	11H00-11H00	11H00-11H00	11H00-11H00	11H00-11H00
Prélevé le		28/08/1997 à 11H28	29/08/1997 à 11H29	30/08/1997 à 11H30	28/08/1997 à 11H00	28/06/1997 à 11H00	28/08/1997 à 11H00
Traitements		J, 1, 2, Z	J	S	C	Z	B, P
Point de prélèvement No		BOUE ACTIV 1	BOUE RECIRC. 1	BOUE RECIRC. 1	BOUE ACTIV 2	BOUE RECIRC. 2	BOUE RECIRC. 2
PHYSICO-CHIMIE							
Mat. suspension	Résultat :	*T 3550	* 4400	* 8930	* 4210	* 4110	* 9310
Norme : >3000	Unité :	mg/l	mg/l	mg/l	mg/l	mg/l	mg/l
	Méthode :	NF T90-105	NF T90-105	NF T90-105	NF T90-105	NF T90-105	NF T90-105
Mat. volatiles	Résultat :	*T 55	* 70	* 77	* 69	* 69	* 52
Norme : De 50 à 100	Unité :	% MS	% MS	% MS	% MS	% MS	% MS
	Méthode :						

SPECIMEN

Fondation reconnue
  d'utilité publique

1, rue du Professeur Calmette
B.P. 245 - 59019 Lille cedex
Tél. 03 20 87 77 30 à 33 - Fax 03 20 87 73 83

Laboratoire de référence agree
pour l'analyse des eaux

ANALYSERAPPORT

Blad 1 van 3

opdrachtgever	datum en kenmerk opdracht
Kiwa N.V. opdrachtnummer	Memo; 10-11-1997 analysedatum/ -periode
30.1263.017; 97-206	12-11-1997

Onderzoek en Advies

Kiwa N.V.
Groningenhaven 7
Postbus 1072
3430 BB Nieuwegein
Telefoon (030) 606 95 11
Telefax (030) 606 11 65
E-mail alg@kiwaoa.nl
Internet www.kiwa.nl

Resultaten

Parameters	Monster 1
Bacteriën van de coligroep, totaal KVE/300 ml	< 1
Bacteriën van de coligroep, thermotolerant KVE/300 ml	< 1
Sporen sulfiet reducerende Clostridia KVE/100 ml	< 1
Faecale streptococcen KVE/100 ml	< 1
Koloniegetal 22°C KVE/ml	12
Koloniegetal 37°C KVE/ml	2

Toelichting bij de resultaten

Kodering opdrachtgever: Monster 1
Monstercodenummer(s) : M-971494
Datum monsters   : 11-11-1997

parameter   : Monsterneming van water
voorschrift(en)   : Huisvoorschrift LMB-018
sterlab-erkend   : ja

parameter   : Bacteriën van de coligroep
                        in water
voorschrift(en)   : Huisvoorschrift LMB-028
techniek   : Membraanfiltratie
aantoonbaarheidsgrens : 1 KVE/ 300 ml
standaardafwijking van de
binnenlabreproduceerbaarheid: 20%
sterlab-erkend   : Ja

Handelsregister
's-Gravenhage, nr. 27039108

ANALYSERAPPORT

Blad 2 van 3

opdrachtgever	datum en kenmerk opdracht	Onderzoek en Advies

Kiwa N.V.
opdrachtnummer

Memo; 10-11-1997
analysedatum/ -periode

Kiwa N.V.
Groningenhaven 7
Postbus 1072
3430 BB Nieuwegein
Telefoon (030) 606 95 11
Telefax (030) 606 11 65
E-mail alg@kiwaoa.nl
Internet www.kiwa.nl

30.1263.017; 97-206

12-11-1997

Toelichting bij de resultaten

parameter   : Thermotolerante bacteriën van de
                                               coligroep in water
voorschrift(en)   : Huisvoorschrift LMB-028
techniek   : Membraanfiltratie
aantoonbaarheidsgrens   : 1 KVE/ 300 ml
standaardafwijking van de
binnenlabreproduceerbaarheid: 25%
sterlab-erkend   : Ja

parameter   : Faecale streptococcen in water
voorschrift(en)   : Huisvoorschrift LMB-029
techniek   : Membraanfiltratie
aantoonbaarheidsgrens   : 1 KVE/ 100 ml
standaardafwijking van de
binnenlabreproduceerbaarheid: 20%
sterlab-erkend   : Ja

parameter   : Sporen van sulfiet reducerende
                                               Clostridia in water
voorschrift(en)   : Huisvoorschrift LMB-033
techniek   : Membraanfiltratie
aantoonbaarheidsgrens   : 1 KVE/ 100 ml
standaardafwijking van de
binnenlabreproduceerbaarheid: 25%
sterlab-erkend   : Ja

ANALYSERAPPORT

Blad 3 van 3

		Onderzoek en Advies
opdrachtgever	datum en kenmerk opdracht	Kiwa N.V.
Kiwa N.V.	Memo; 10-11-1997	Groningenhaven 7 Postbus 1072
opdrachtnummer	analysedatum/ -periode	3430 BB Nieuwegein Telefoon (030) 606 95 11
		Telefax (030) 606 11 65
30.1263.017; 97-206	12-11-1997	E-mail alg@kiwaoa.nl Internet www.kiwa.nl

Toelichting bij de resultaten

```
parameter               : Koloniegetal bij 22°C in water
voorschrift(en)         : Huisvoorschrift LMB-032
techniek                : Gietplaat
aantoonbaarheidsgrens   : 1 KVE/ ml
herhaalbaarheid         : <30 KVE <100%
                          >30 KVE < 50%
sterlab-erkend          : Ja
```

```
parameter               : Koloniegetal bij 37°C in water
voorschrift(en)         : Huisvoorschrift LMB-032
techniek                : Gietplaat
aantoonbaarheidsgrens   : 1 KVE/ ml
herhaalbaarheid         : <30 KVE <100%
                          >30 KVE < 50%
sterlab-erkend          : Ja
```

Opmerkingen: geen

Handelsregister
's-Gravenhage, nr. 27039108

Chapter 10

Accreditation

10.1. Introduction

The implementation of a quality assurance programme will give confidence in the validity of the results produced. The ability to make meaningful decisions on compliance with specifications and enforcement of legislation requires third-party assessment to ensure consistent reliability. This third-party assessment of the processes a laboratory implements to achieve consistent quality is known as accreditation. It involves assessment against standards which apply to all laboratories involved in a wide range of measurement and testing. The application of these standards to laboratories involved in food and water microbiological testing is at an early stage of development but target standards have been set in many countries.

10.2. International aspects of accreditation

Most national laboratory accreditation schemes throughout the world have based their accreditation criteria on ISO Guide 25 'General requirements for the competence of calibration and testing laboratories'. Countries will often have different measurement and testing infrastructures and each country has produced its own criteria. In Europe, however, in meeting the challenge of the single market the series of standards EN45000 has been published. These set down the requirements for the operation of accreditation and certification bodies. In Europe, all EU countries have set up their own accreditation schemes. These, and other schemes joined together in 1989 to form the Western European laboratory accreditation cooperation (WELAC), and the nationally recognized accreditation bodies which have signed the WELAC memorandum of understanding are from the following member states (note: WELAC and WEEC have now merged, and operate as EAL):

– Austria
– Belgium
– Denmark
– Finland
– France
– Germany
– Greece
– Iceland
– Ireland
– Italy
– Netherlands
– Norway
– Portugal
– Spain
– Sweden
– Switzerland
– United Kingdom

WELAC cooperates with a number of other organizations in Europe which have an interest in laboratory accreditation including:

– EURACHEM
– OECD
– WECC (Western European Calibration Cooperation) and Eurolab.

10.3. The European single market

One of the main aims of the European single market is that goods and services should be able to cross national boundaries without the need for retesting as this can be both costly and time-consuming. The achieve this, any testing carried out in any EU Member State should automatically be accepted in any other EU Member State. In order for this concept to be accepted, all users of laboratories and governments need to be assured that laboratories performing these tests are reliable and working to the same standards. Multilateral agreements have been agreed and progress is being made to achieve common standards.

10.4. National accreditation organizations

The organizations involved in the Western European laboratory accreditation cooperation (WELAC) are given in the following table:

AUSTRIA
Bundesministerium für
wirtschaftliche Angelegenheiten
Haupstrasse, 55–57
1031 Wien
Austria
Tel : +43 1 71102249
Fax : +43 1 7143582

BELGIUM
Organisation Belge des Laboratoires
d'Essais (BELTEST)
Ministerie van Ekonomische Zaken
Central Laboratorium
Rue de la Senne 17A
Bruxelles 1000
Belgium
Tel : +32 2 5117769
Fax : +32 2 5144756

DENMARK
Danish Accreditation Service
(DANAK)
National Agency of Industry & Trade
137, Tagensvej
DK-2200 Copenhagen
Denmark
Tel :+45 45 931144
Fax:+45 31 817068

FINLAND
Centre for Metrology and
Accreditation (FINAS)
PO Box 239 (Lonnrotinkatu 37)
SF-00181 Helsinki
Finland
Tel :+358 0 61671
Fax :+358 0 6167467

FRANCE
COFRAC
Secretariat permanent
77 rue du Père Corentin
75014 Paris
France
Tel :+33 1 40 52 05 70
Fax :+33 1 40 52 05 71

GERMANY
Deutscher Akkreditierungs Rat (DAR)
Bundesanstalt für Materilaforschung
und -prufung (BAM)
Unter den Eichen 87
D-1000 Berlin 45
Germany
Tel :+49 30 81041710
Fax :+49 30 81041717

GREECE
Hellenic Organization for
Standardization (ELOT)
313 Acharnon Street
GR-111 45, Athens
Greece
Tel : +30 1 2022345
Fax : +30 1 2020776

ICELAND
Iceland Bureau of Legal Metrology
PO Box 8114
IS-128 Reykjavik
Iceland
Tel : +354 1 681122
Fax : +354 1 685998

IRELAND
Irish Certification and Laboratory
Accreditation Board (ICLAB)
Glasnevin
Dublin 9
Ireland
Tel : +353 1 8370101
Fax : +353 1 8368738

ITALY
SINAL
Via Campania 31
00187 Roma
Italy
Tel : +39 6 4871176
Fax : +39 6 4814563

NETHERLANDS
NKO/STERIN/STERLAB
PO Box 29152
3001 GD
Rotterdam
The Netherlands
Tel : +31 10 4136011
Fax : +31 10 4133557

NORWAY
National Accreditation Norway
Direktoratet for Maleteknikk
PO Box 6832, St Olavs Plass
0130 Oslo 1
Norway
Tel : +47 2 2200226
Fax : +47 2 2207772

PORTUGAL
Instituto Portugues de Qualidade
(IPQ)
Rue Jose Estavao 83A
1199 Lisboa codes
Portugal
Tel : +351 1 523978
Fax : +351 1 3530033

SPAIN
Red Espanola de Laboratoires de
Ensayo (RELE)
Avda. De Concha Espina, 65,2
28016 Madrid
Spain
Tel : +34 1 5649687
Fax : +34 1 5630454

SWEDEN
SWEDAC
PO Box 878
S-501 15 Boras
Sweden
Tel : +46 33 177700
Fax : +46 33 101392

SWITZERLAND
Swiss Federal Office of Metrology
Lindenweg 50
CH-3084 Wabern
Switzerland
Tel : +41 31 9633412
Fax : +41 31 9633210

UNITED KINGDOM
NAMAS Executive
National Physical Laboratory
Queens Road
Teddington
Middlesex TW11 0LW
United Kingdom
Tel : +44 181 943 6554
Fax : +44 181 943 7134

SECRETARY OF WELAC
NAMAS Executive
National Physical Laboratory
Queens Road
Teddington
Middlesex TW11 0LW
United Kingdom
Tel : +44 181 943 6554
Fax : +44 181 943 7134

10.5. Accreditation requirements

When laboratories are accredited, the quality system must include:

(a)   Organization and management;

(b)   Quality assurance programme;

(c)   Audit and review;

(d)   Staff;

(e)   Equipment;

(f)   Calibration;

(g)   Methods and procedures;

(h)   Environment;

(i)   Sample handling;

(j)   Records;

(k)   Reports;

(l)   Complaints and anomalies;

(m)   Sub-contracting;

(n)   Use of outside support services and supplies .

In addition to these aspects of the criteria which are common to all laboratories irrespective of discipline, there are other requirements which need specific interpretation for food and water microbiological testing. These are as follows.

(i) Training staff

For food and water testing it is especially important for staff to receive adequate and appropriate training and to have the opportunity to be involved in a range of different sampling occasions and types in order to build up their experience.

(ii) Quality control

Analytical performance should be monitored by operating quality control schemes which are appropriate to the type and frequency of testing undertaken by a laboratory. The range of quality control activities available to laboratories includes the use of:

(a)   certified reference materials (CRMs);

(b)   secondary reference materials (RMs);

(c)   spiked samples, standard additions and internal standards;

(d)   replicate testing;

(e)   alternative validated methods;

(f)   control charts.

An effective means by which a laboratory can monitor its performance is by taking part in proficiency testing schemes and interlaboratory comparisons. In some circumstances, accreditation bodies may specify participation in particular proficiency testing schemes as a requirement of accreditation.

It is not so easy to apply some of these criteria to sampling activities, but in these cases staff training and quality auditing assume greater significance.

(iii) Equipment and calibration

In environmental testing laboratories, sampling containers and glassware often require rigorous cleaning programmes to overcome potential problems related to cross-contamination and adsorption interferences. The need to calibrate equipment is generally well understood but establishing traceability to national standards, where this is possible, is less well known.

(iv) Reference materials

In all analyses, but particularly in environmental analysis, it is important to select reference materials which have a matrix as close as possible to the samples under analysis, where these are available.

(v) Methods and procedures

Wherever possible and appropriate, pre-validated standard or reference methods should be used. These should include the necessary provision for controls and blanks. Validation of methods should include, but not be limited to, the demonstration of sensitivity, selectivity and precision. Sampling and analysis methods for water are in many cases well established and standardized.

(vi) Environment

It is important that laboratory conditions should be adequate to prevent cross-contamination between samples of different analyte concentrations and contamination from the laboratory environment.

10.6. Achieving accreditation

It will be seen that the introduction of a quality assurance programme based on the procedures described in the earlier chapters of this book and which are appropriate to the scope of testing of the laboratory will form a sound basis for accreditation. Not everything described in the earlier chapters will be necessary for all laboratories, and quality managers/coordinators will be able to choose those elements of the QA programme which, in their judgement, are appropriate to their own laboratory. Of course, all these elements must be documented and will form the quality manual. Having made progress as described in Chapter 2 (Implementation of QA programmes), the laboratory should then apply to its own national accreditation body for further guidance and help with understanding the third-party setting, and assessment of compliance with standards. Their accreditation body publications will give help and guidance so that an application for accreditation can be made. Following the application and payment of the necessary fee, an initial visit will be made by the assessors of the accreditation body. Usually, there will be minor modifications to laboratory procedures required and justification for any unusual methods, often requiring validation data. Once these are resolved, the laboratory will be accredited for those analyses assessed and will be subject to further visits at intervals to ensure that standards are being maintained. Eurachem have just published the 'Accreditation Guide for Laboratories Performing Microbiological Testing' which gives guidance for the interpretation of EN45001 and ISO Guide 25.

Annex

Statistics for quality assurance in food and water microbiology

The validation of microbiological methods and the interpretation of their results is complicated in that these often demonstrate a certain degree of variability. This variability can be ascribed to systematic effects as well as to random variation connected with sampling and the performance of analytical procedures. To take account of this the collection of multiple information is required and is achieved by repeated counting, parallel plating and multiple procedures. In such cases the single results can be summarized to provide an evaluation that describes the main features of the analysis and reduces the influence of random variability.

Important statistical methods have been developed to
- summarize the main characteristics of multiple samples and to estimate the influence of random variation
- generalize the sample results for the population or environment from which they were taken
- compare the results from different samples taking account of random variation so that systematic and random differences can be distinguished.

1. Some basic calculations

Two main characteristics of a set of observations — for instance a set of parallel plate counts — are its location and its dispersion. In the case of quantitative measurements such as counts, volumes or concentrations the statistics often used are the arithmetic mean to describe the location, and the variance, the standard deviation or the coefficient of variation as measures of dispersion.

Arithmetic mean

The *arithmetic mean* or *average* $\bar{x}$ is defined as the sum of all single measurements x_i divided by the number of values n.

$$\bar{x} = \frac{1}{n} \sum_{i=1}^{n} x_i$$

However this location statistic gives only the right impression of the data set if the single values are distributed symmetrically around the mean value as each measurement makes the same contribution to the average. For skewed distributions the influence of the highest (or lowest) values would yield averages lying apart from the majority of observations.

Variance, standard deviation and coefficient of variation

The corresponding measure of dispersion is the *variance* s^2 which is calculated using the differences between the single measurements x_i and their average $\bar{x}$. These differences are squared to ignore the direction of deviation, they are summed up, and then the sum is divided by the number of measurements minus one.

$$s^2 = \frac{1}{n-1} \sum_{i=1}^{n} (x_i - \bar{x})^2$$

An alternative formula yielding the same result is

$$s^2 = \frac{1}{n-1}\left(\sum_{i=1}^{n} x_i^2 - \frac{1}{n}\left(\sum_{i=1}^{n} x_i\right)^2\right)$$

The first formula, however, is easier to understand as it comes close to the average squared difference of measured values from their arithmetic mean.

Hint: Sometimes this average squared difference of measured values can be found as the variance of a data set. In such cases the variance s^2 is defined by dividing the sum of squared differences simply by n instead of $(n - 1)$. To avoid different results from the use of different definitions, the formula used for calculation should always be checked. These problems may arise especially when using computer programs for statistical analyses.

One disadvantage of describing the variation within a data set in terms of the variance however is that this value is not so easy to interpret because of the squared deviations used in the calculation. Therefore the *standard deviation s* is often given which is gained by taking the square root of the variance.

$$s = \sqrt{s^2} = \sqrt{\frac{1}{n-1}\sum_{i=1}^{n}(x_i - \bar{x})^2}$$

A comparison of the variation in two or more sets of measurements with different means should be based on a third deviation statistic that is calculated from the standard deviation by dividing it by the mean. This one is called the *coefficient of variation* or the *relative standard deviation*: $v = s/\bar{x}$. It gives the variation of the measurements in relation to their average.

Example:

Assume a set of six parallel plates was incubated and the colonies counted. The $n = 6$ counting results are:

$x_1 = 86, x_2 = 113, x_3 = 92, x_4 = 98, x_5 = 87, x_6 = 94$

From these data the following location and dispersion statistics can be calculated:

$\bar{x} =$ 95.00

$s_2 =$ 97.60

$s =$ 9.88

$v =$ 0.10

Note 1: It has already been mentioned that the use of arithmetic mean and variance or standard deviation is only appropriate in the case that the measurements show a symmetrical distribution. Bacterial counts however tend to show skewed distributions to the right because of some higher counting results among more lower ones which cannot be lower than zero. In such situations the *geometric mean* is more appropriate than the arithmetic mean as it is always less or equal $\bar{x}$.

The geometric mean $\bar{x}_g$ is defined as the nth root drawn of the product of all measurements.

$$\bar{x}_g = \sqrt[n]{x_1 \cdot x_2 \cdot \ldots \cdot x_n}$$

This value is closely related to the arithmetic mean on the logarithmic scale as one would get $\bar{x}_g$ as well by taking the arithmetic mean of the log-transformed values to the chosen base, for instance

$$x_g = 10^{\overline{\lg x}}$$

For the colony counts from the example a geometric mean of $\bar{x}_g =$ 94.50 (with $\overline{\lg x} = 1.976$) can be calculated.

Median

A different way to deal with data that are not symmetrically distributed would be to give the *median* as a location statistic. The first step in deriving this value is to generate the ordered list of values and to denote for each value the position number or rank it has in this list (these are the numbers 1,...,n). The median $\widetilde{x}$ is defined as that value ranking at position $(n + 1)/2$ in case n is an odd number:

$$\widetilde{x} = x_{[(n+1)/2]}$$

or it is the average of the two values ranking on positions $(n/2)$ and $(n + 2)/2$ in case n is an even number:

$$\widetilde{x} = (x_{[n/2]} + x_{[(n+2)/2]}) / 2$$

As the median separates the lower half of measurements and the higher half of them, it is called the 50%-point $x_{0.50}$ as well.

For the example data set the ordered list of values is:

$$x_{[1]} = 86, x_{[2]} = 87, x_{[3]} = 92, x_{[4]} = 94, x_{[5]} = 98, x_{[6]} = 113$$

and the median is:

$$\widetilde{x} = (x_{[3]} + x_{[4]}) / 2 = \frac{92 + 94}{2} = 93$$

Note 2: A different problem arises if the single results — for instance colony counts — are not measured for the same concentration of sample material but for more than one countable dilution. In such a situation the mean count should not be calculated as the simple arithmetic mean because that would give equal weights to all counts no matter which amount of sample material they represent. Instead the mean count should be obtained as a *weighted average*

$$\widetilde{x}_{\mathrm{w}} = \frac{\sum \sum x_{ij}}{\sum \sum a^{-1} \cdot v_{ij}}$$

where x_{ij} is the colony count of the jth plate of the ith dilution, v_{ij} is the plating volume of the jth plate of the ith dilution, and a is the dilution factor between steps.

With this calculation the total number of colonies is not related to the number of plates but to the total amount of sample material analysed.

Example 1:

Assume there are measurements from two dilutions 1:10 and 1:100 from which three plates were incubated with a plating volume of 1 ml and the colonies were counted.

1:10 $x_{11} = 103$ $x_{12} = 92$ $x_{13} = 99$

1:100 $x_{21} = $ 8 $x_{22} = 12$ $x_{23} = $ 9

The sum of colony counts is 323

The plating volume on each plate is $v_{ij} = 1$. The dilution factor a is given as 0.1 so that the amount of sample material per plate is 0.1 ml for the first dilution and 0.01 ml for the second. Then the total amount of sample material the plates represent is $3 \cdot 0.1 + 3 \cdot 0.01 = 0.33$ ml. From this one gets a weighted average for the colony count of $323/0.33 = 978.79$ per 1 ml of sample material.

Example 2:

Assume there are measurements from five dilutions 1:2, 1:4, 1:8, 1:16, 1:32 from which three plates were analysed giving the colony counts:

1:2 $x_{11} = 204$ $x_{12} = 198$ $x_{13} = 198$

1:4 $x_{21} = 103$ $x_{22} = 111$ $x_{23} = 104$

1:8 $x_{31} = $ 53 $x_{32} = $ 45 $x_{33} = $ 46

1:16 $x_{41} = $ 25 $x_{42} = $ 31 $x_{43} = $ 22

1:32 $x_{51} = $ 10 $x_{52} = $ 14 $x_{53} = $ 12

The sum of colony counts is 1176.

The plating volume on each plate is $v_{ij} = 1$. The dilution factor a is given as 0.5 so that the amount of sample material is 0.5 for the first dilution, 0.25 for the second, 0.125 for the third and 0.0625 and 0.03125 for the fourth and fifth respectively. Then the total amount of sample material the plates represent is $3 \cdot 0.5 + 3 \cdot 0.25 + 3 \cdot 0.125 + 3 \cdot 0.0625 + 3 \cdot 0.03125 = 2.90625$ ml. From this a weighted average for the colony count of 404.65 per 1 ml of sample material is calculated.

2. Some basic distributions

Many statistical procedures deal with the problem of how to generalize sample results and how to draw conclusions from sample comparisons. These methods are based on assumptions or models which describe the population or the environment the samples are taken from and which define the probabilities that certain sample results will occur. By using such a statistical model it is possible to describe for instance the random variability of results when sampling from a well mixed suspension, where `homogenization' is assumed and to calculate the probability that they will lie within a given range. (When speaking of sampling it should be stressed that in statistical terms the expression 'sample size' refers to the number of single samples, sample units or 'sub samples' that are taken, not to the size of one single sample. Here sample sizes are usually denoted by n.)

Some basic distribution models are the Bernoulli and the binomial distribution, the normal and the log-normal distribution and the Poisson distribution.

Bernoulli distribution

A Bernoulli trial is one single random trial with only two possible outcomes like 'success' or 'failure', 'positive' or 'negative', 'present' or 'absent'. The probability that a single measurement of this kind will give a positive result is usually denoted by p, accordingly the probability of a negative result is $1 - p$.

Example:

Taking a single random sample from a well mixed batch with 90% salmonella positive and 10% salmonella negative units the probability to pick out a positive sample is $p = 0.90$.

Binomial distribution

If a Bernoulli trial is repeated several (n) times independently from each other, so that for each repetition the probability of a positive outcome is p, and if one is interested in the number of positive outcomes among these n trials, then this is a situation in which a binomial distribution can be applied. This distribution model gives a formula to calculate the probability $P(x)$ to get exactly a number of x positive outcomes when the trial is repeated n times.

$$P(x) = \binom{n}{x} p^x (1-p)^{(n-x)}$$

where $\binom{n}{x}$ is the binomial coefficient.

$$\binom{n}{x} = \frac{n!}{(n-x)!\,x!}$$

with $n! = n \cdot (n-1) \cdot (n-2) \cdot \ldots \cdot 2 \cdot 1$. This coefficient gives the number of different combinations in which x positive and $(n-x)$ negative outcomes can occur among n trials. Probability values for the binomial distribution are tabulated for various combinations of the parameters n and p and are nowadays mainly calculated using computer programs.

Example:

Assume a number of $n = 10$ test tubes are incubated with random subsamples from a sample material so that for each single tube the probability that it gives a positive result is $p = 0.9$. Then, using the binomial distribution, the probabilities that no tube, that exactly one tube, exactly two tubes will give a positive result, and so on, can be calculated. Therefore this distribution provides a basic model to evaluate MPN results.

Normal distribution

This distribution model is based on more complex mathematical considerations, but its use is often motivated by its empirical

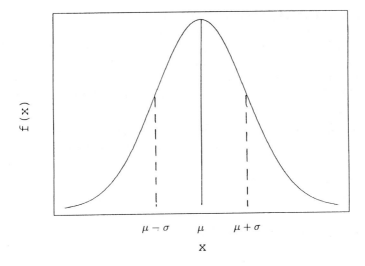

Fig. 1. Normal distribution curve.

evidence. For instance, volume measurements or calibration variabilities are often described as normally distributed.

Figure 1 shows its typical bell shape which is characterized by its symmetry. The location parameter μ gives the expected value or mean, the deviation parameter σ denotes the standard deviation and expresses the spread of the bell.

The possible measurements are marked at the horizontal axis. The probability $P(X \leq a)$ to find a measurement value below a certain limit a is to be derived as the area under the bell shaped curve left to this limit. Accordingly, the probability to get a value lying in a certain interval $[a; b]$ on this axis is calculated as the area between the interval borders a and b or as $P(X \leq b) - P(X \leq a)$, respectively.

As these probability values are quite difficult to obtain they are tabulated as well, but only for the so-called standard normal distribution which has as parameters a mean value of 0 and a standard deviation of 1. All other normal distributions for any combination of μ and σ are related to the standard normal distribution by a relatively simple formula. Assume having measurements which are normally distributed with mean μ and standard deviation σ. Then the probability that a single randomly taken value will not exceed a specified value $x (P(X \leq x))$ equals the probability that a standard normal variate will not exceed a transformed value z, $(P(X \leq x) = P(Z \leq z))$. This value z is derived with the transformation formula:

$$z = \frac{x - \mu}{\sigma}$$

Calculating this transformed value z and looking up tabulated probabilities for the standard normal distribution or having them calculated by computer programs thus provide convenient ways to work with this distribution model.

Some special probabilities are those for central intervals lying symmetrically around the mean μ. Normally distributed measurements have a probability of 68.27% to lie between $\mu - \sigma$ and $\mu + \sigma$, one of 95.45% to lie between $\mu - 2 \cdot \sigma$ and $\mu + 2 \cdot \sigma$ and one of 99.73% to lie between $\mu - 3 \cdot \sigma$ and $\mu + 3 \cdot \sigma$. This characteristic is the basis for designing control charts with theoretical warning limits at $\mu \pm 2 \cdot \sigma$ and action limits at $\mu \pm 3 \cdot \sigma$.

When the real mean μ and the real standard deviation σ in the batch or population are unknown, they have to be estimated. This can be achieved by taking some random samples, analysing them and calculating their arithmetic mean $\bar{x}$ as an estimate for μ and their standard deviation s as *an* estimate for σ.

Example:

Assume 5 ml-pipettes bought from a certain manufacturer are known to have calibration marks that are normally distributed with mean $\mu = 5$ ml and a standard deviation of $\sigma = 0.1$ ml (2% of the mean) as is shown in Fig. 2. Then the probability that a pipette randomly taken out of this batch has a calibration mark at or below 4.8 ml is

$$P(X \leq 4.8) = P(Z \leq (4.8 - 5) / 0.1) = P(Z \leq -2) = 0.0228$$

The probability to take out a pipette having its calibration mark between 4.9 and 5.1 is

$$P(4.9 \leq X \leq 5.1) = P(X \leq 5.1) - P(X \leq 4.9) = P(Z \leq 1) - P(Z \leq -1)$$

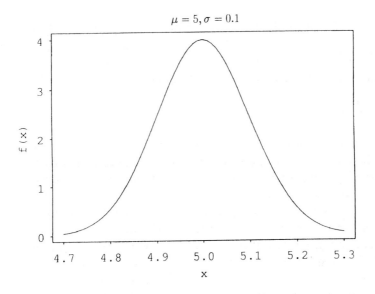

Fig. 2. Normal distribution with μ = 5, σ = 0.1.

or the area marked in Fig. 3, respectively. Note that the last interval is the same as $[\mu - \sigma \,; \mu + \sigma]$ and thus the probability is 0.6827.

Log-normal distribution

Bacterial counts are often characterized by a skew distribution of relatively more low then high values. One of the various distribution models describing skewed shapes is the log-normal distribution. By taking logarithms of measurement data values are obtained that are again normally distributed.

Example:

See Section 8.2.6; these control charts are based on log-transformed values, the original log-normal distribution and the normal distribution for the transformed values are shown in Fig. 4.

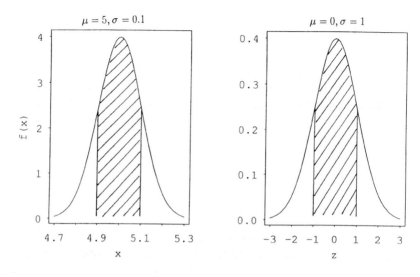

Fig. 3. Normal distribution for example data and standard normal
distribution.

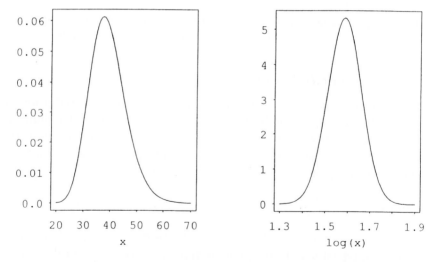

Fig. 4. Log-normal and normal distribution in case the log-transformed values
have a mean of 1.583 and a standard deviation of 0.072.

Poisson distribution

The Poisson distribution is another model with great importance for statistical analyses of microbiological data. If micro-organisms are randomly distributed in the sample concerned and do not affect each other's position in the sample, bacterial counts are Poisson distributed. The probability to find x units in a randomly taken sample is given by the formula:

$$P(x) = e^{-\lambda}\, \frac{\lambda^x}{x!}$$

As this distribution describes only the unavoidable random variation connected with bacterial counts, samples following this model are sometimes called 'well homogenized'.

Figure 5 shows Poisson distributions with different values for λ. The Poisson distribution is characterized by its parameter λ which is the mean or expected value and at the same time its variance. So if bacterial counts show a mean value of 100 units their variance is 100

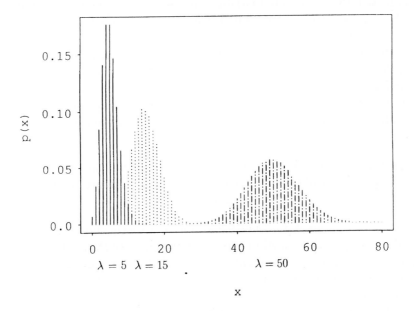

Fig. 5. Poisson distributions.

units as well and their standard deviation is 10 units. The shape of the Poisson distribution is slightly skewed to the right for low mean values, but it becomes more and more symmetrical the bigger the mean values are. If the true mean value of Poisson distributed measurements is unknown it can be estimated from samples by calculating their arithmetic mean $\bar{x}$.

Index of dispersion

As the Poisson distribution can be regarded as the model for 'homogeneity' (in the statistical sense of the word) or for plating from a well mixed sample, a variability of plate counts according to this model is to be interpreted as simply random. Only a dispersion bigger than the Poisson variability could be an overdispersion giving hints to incomplete homogenization, performance errors, etc. It should be mentioned however that natural samples are often overdispersed for various reasons. Therefore a check on overdispersion is useful to decide whether Poisson distribution can be assumed for further analyses or whether some kind of transformation of values should be applied first.

A first check on overdispersion for a set of bacterial counts from parallel plates could be to calculate their average $\bar{x}$ and their variance s^2 and compare them. If the Poisson model can be assumed, they should yield similar values ($\bar{x} \approx s^2$), but some differences can be expected nonetheless as the values are only calculated from sample results and thus will vary themselves.

Hint: If samples are taken from a Poisson distribution the natural logarithms of the results should show a standard deviation near to 0.10; this is used in Chapters 7 and 8).

Therefore to check for overdispersion a statistic should be calculated so that the distribution can be derived in case that samples are taken from a Poisson distribution. This distribution information is necessary to get critical limits and to distinguish between random overdispersion of the samples and a substantial one.

Such a dispersion statistic for n single measurements x_i is the index of dispersion D^2:

$$D^2 = \sum_{i=1}^{n} \frac{(x_i - \bar{x})^2}{\bar{x}} = \frac{n \sum x_i^2 - (\sum x_i)^2}{\sum x_i}$$

A related alternative dispersion statistic is the log-likelihood statistic G^2:

$$G^2 = 2 \cdot \sum_{i=1}^{n} x_i \ln \frac{x_i}{\bar{x}} = 2 \cdot \sum_i x_i \ln x_i - 2 \cdot \sum_i x_i \ln \frac{\sum x_i}{n}$$

For measurements following a Poisson distribution both statistics, D^2 and G^2, will adopt a χ^2-distribution with $(n-1)$ degrees of freedom. For this distribution the number of degrees of freedom $(n-1)$ gives the expected value for D^2 or G^2, respectively.

Hint: Both these are the appropriate formulas in the case that all single values have the same weight for the calculation, for instance for the evaluation of parallel plate counts with equal volumes on all plates. In different contexts special formulas are to be used which are referred to in Chapter 8.

Example:
For the set of six parallel plates giving the $n = 6$ counting results:

86, 113, 92, 98, 87, 94

the average $\bar{x} = 95$ and the variance $s^2 = 97.6$ are quite similar yielding a ratio $s^2/\bar{x}$ of 1.03. The dispersion index is $D^2 = 5.137$, calculating the log-likelihood statistic gives a value of $G^2 = 4.988$, for both statistics the degrees of freedom are $n - 1 = 5$.

3. Confidence intervals

In the last paragraph some location and deviation statistics are mentioned as suitable estimates for the parameters of distribution models, for instance the average $\bar{x}$ to estimate the mean value μ of a normal distribution and to estimate the mean value λ of a Poisson distribution, or the empirical variance s^2 to estimate the variance σ^2

of a normal distribution. This method is called point estimation because the estimate is a single point value calculated from a sample.

However, even different sets of samples taken from the sample population or batch will give different values for the location and deviation statistics. This is a natural characteristic of sample investigations, only the differences will be smaller or bigger depending on the number of sample units taken for each set and on the variability of single results. Consequently point estimates are characterized by a certain variation themselves.

To take account of this variability, interval estimation is often used instead. The idea behind this procedure is to derive borders a and b so that the interval $[a;b]$ will cover the unknown value or population parameter θ with a given probability $(1 - \alpha)$:

$$P(a \le \theta \le b) = 1 - \alpha$$

The probability $(1 - \alpha)$ is usually called the confidence level of the interval estimation, therefore these intervals are known as *confidence intervals*.

Example 1:

The mean value μ of calibration marks on pipettes should be estimated for which a normal distribution with a standard deviation of $\sigma = 0.1$ ml can be assumed. A sample of $n = 10$ pipettes is taken at random and their volumes are checked yielding an average filling of 4.98 ml.

In general, arithmetic means of values drawn from normal distributions are normally distributed themselves with the same expected value μ as for single values, but with a smaller standard deviation. This can be derived as $\sigma_{\bar{x}} = \sigma / \sqrt{n}$ which is depending on the standard deviation σ of single values and on the number of sample units n. In this case the following relationship holds:

$$P\left(\overline{X} - z_{1-\frac{\alpha}{2}} \cdot \frac{\sigma}{\sqrt{n}} \le \mu \le \overline{X} + z_{1-\frac{\alpha}{2}} \cdot \frac{\sigma}{\sqrt{n}} \right) = 1 - \alpha$$

giving the formula for the interval as:

$$\left[\overline{X} - z_{1-\frac{\alpha}{2}} \cdot \frac{\sigma}{\sqrt{n}} \; ; \; \overline{X} + z_{1-\frac{\alpha}{2}} \cdot \frac{\sigma}{\sqrt{n}} \right]$$

for a given confidence level $(1 - \sigma)$.

The corresponding $z_{1-\frac{\alpha}{2}}$ values for some common confidence levels are:

$1 - \alpha$	α	$z_{1-\frac{\alpha}{2}}$
0.90	0.10	1.645
0.95	0.05	1.960
0.99	0.01	2.576

For the example, the 90%, 95% and 99% confidence intervals for μ are calculated as

$$\left[4.98 \pm 1.645 \cdot \frac{0.1}{\sqrt{10}} \right] = [4.928; 5.032]$$

$$\left[4.98 \pm 1.960 \cdot \frac{0.1}{\sqrt{10}} \right] = [4.918; 5.042]$$

$$\left[4.98 \pm 2.571 \cdot \frac{0.1}{\sqrt{10}} \right] = [4.899; 5.061]$$

Example 2:

The mean value μ of measurements should be estimated; for these a normal distribution can be assumed, but the standard deviation is unknown as well.

In this case has to be estimated by the empirical standard deviation s from the same sample that is taken to estimate μ. Consequently, arithmetic averages will not follow a normal distribution but instead a t-distribution with $(n - 1)$ degrees of freedom. The resulting formula for confidence intervals then is:

$$\left[\overline{X} - t_{n-1;\left(1-\frac{\alpha}{2}\right)} \cdot \frac{s}{\sqrt{n}} ; \overline{X} + t_{n-1;\left(1-\frac{\alpha}{2}\right)} \cdot \frac{s}{\sqrt{n}}\right]$$

(*t*-values for different combinations of α and degrees of freedom can be found in statistical tables.)

Assume the $n = 10$ pipettes in the example show a variation of $s = 0.09$ ml. The 95% confidence interval for μ is calculated, with

$$t_{n-1=9;\left(1-\frac{\alpha}{2}\right)=0.975} = 2.262$$

as

$$\left[4.98 - 2.262 \cdot \frac{0.09}{\sqrt{10}} ; 4.98 + 2.262 \cdot \frac{0.09}{\sqrt{10}}\right] = [4.9156; 5.0444]$$

Example 3:

The parameter p of a binomial distribution should be estimated from a sample, for instance to estimate the proportion of false negative results for some presence/absence test. A trial with $n = 100$ repetitions is carried out using samples that are known to be positive. These n replications yield a negative result 25 times, and the correct positive outcome 75 times. Then the point estimate for p is $25/100 = 0.25 = \hat{p}$.

As the number of trials is quite large ($n \geq 100$), the point estimate $\hat{p}$ is approximately normally distributed with variance $\hat{p} \cdot (1 - \hat{p})$. From this the following formula for a confidence interval can be derived:

$$\left[\hat{p} - z_{1-\frac{\alpha}{2}} \cdot \sqrt{\frac{\hat{p} \cdot (1 - \hat{p})}{n}} ; \hat{p} + z_{1-\frac{\alpha}{2}} \cdot \sqrt{\frac{\hat{p} \cdot (1 - \hat{p})}{n}}\right]$$

For the example, the variance is estimated as $\hat{p} \cdot (1 - \hat{p}) = 0.25 \cdot 0.75 = 0.1875$, hence the 95% confidence interval is

$$\left[0.25 - 1.96 \cdot \sqrt{\frac{0.1875}{100}} ; 0.25 + 1.96 \cdot \sqrt{\frac{0.1875}{100}}\right] = [0.165; 0.335]$$

4. Statistical tests

The aim of a statistical test is to decide whether sample results — often gained in planned experimental studies — are in accordance with theoretical considerations or hypotheses or whether they give evidence against these. The main steps in a statistical test procedure are:
- to formulate the statistical hypothesis to be tested, this is usually called the null-hypothesis H_0, and the alternative hypothesis H_1 it is tested against;
- to decide on the number of sample units to be investigated to come to a decision;
- to decide on the so-called significance level α for the test; this is the probability to come to a rejection of the null-hypothesis in case it is really true;
- to decide on a test statistic which will be calculated from the sample results and to derive its critical values according to the chosen significance level; with this step a decision rule is defined;
- then finally to perform the trial or the data collection, to calculate the test statistic and to make the decision according to the rule.

Example 1:

Assume a new batch of 5 ml-pipettes is bought for which it is known that the calibration marks are normally distributed with a standard deviation of $\sigma = 0.1$ ml. Now it should be tested whether the expected mean is 5 ml or whether there is a systematic bias giving too small or too high volumes.
- The statistical hypothesis to be tested would be H_0: $\mu = \mu_0 = 5$ ml against H_1: $\mu \neq 5$ ml.
- A sample size of $n = 10$ is chosen.
- The significance level $\alpha \leq 0.05 = 5\%$ is fixed.
- The appropriate test statistic for this problem is the arithmetic mean of the sample results because this is a good estimate for the mean of the batch the samples are taken from.

In case H_0 is really true, it can be derived that the sample average $\bar{x}$ will be lying within the range

$$\left[\mu_0 - z_{1-\frac{\alpha}{2}} \cdot \frac{\sigma}{\sqrt{n}} ; \mu_0 + z_{1-\frac{\alpha}{2}} \cdot \frac{\sigma}{\sqrt{n}}\right]$$

with a probability of $(1 - \alpha) = 0.95$.

For the example this interval is calculated as

$$= 5 \pm 1.96 \cdot \frac{0.1}{\sqrt{10}} = [4.938; 5.062]$$

therefore the decision rule is to reject H_0, if $\bar{x} < 4.938$ or if $\bar{x} > 5.062$, otherwise H_0 cannot be rejected.

Note: An equivalent formulation for the decision rule would be as follows.

Calculate a standardized test statistic z

$$z = \frac{\bar{x} - \mu_0}{\sigma / \sqrt{n}}$$

and compare it with its critical limits $\pm z_{1-\frac{\alpha}{2}}$. If z is lying outside this interval, H_0 is rejected.

For the example, the critical limits for z are ± 1.96. For a sample average of 4.98 the score is $(4.98 - 5)(0.1 / \sqrt{10}) = -0.633$.

Example 2:

Now assume the standard deviation σ is unknown. Then in analogy to the interval estimation mentioned before σ has to be estimated by the empirical standard deviation s from the sample. As a consequence the critical values have to be based on the t- instead of the normal distribution.

- The statistical hypothesis to be tested is again formulated as $H_0: \mu = \mu_0 = 5$ ml against $H_1: \mu$ 5 ml.
- A sample size of $n = 10$ is chosen.
- The significance level $\alpha \le 0.05 = 5\%$ is fixed.
- The appropriate test statistic for this problem is the arithmetic mean of the sample results for the same reasons as before.

In case H_0 is really true and σ is estimated by the empirical standard deviation as $s = 0.09$ it can be derived that the sample average $\bar{x}$ will be lying within the range

$$\left[\mu_0 - t_{n-1;\left(1-\frac{\alpha}{2}\right)} \cdot \frac{\sigma}{\sqrt{n}} ; \mu_0 + t_{n-1;\left(1-\frac{\alpha}{2}\right)} \cdot \frac{\sigma}{\sqrt{n}} \right]$$

with a probability of $(1 - \alpha) = 0.95$.

For the example this interval is calculated as

$$= 5 \pm 2.262 \cdot \frac{0.09}{\sqrt{10}} = [4.9356; 5.0644]$$

therefore the decision rule is to reject H_0, if $\bar{x} < 4.9356$ or if $\bar{x} > 5.0644$, otherwise H_0 cannot be rejected.

Because it is based on the arithmetic means following a t-distribution, this test is called t-test. In the situation described here it is performed with just one sample (of 10 units) and a fixed hypothetical value μ_0 for the mean.

Another common form of the t-test is that for two-sample problems which compares the means of two sample sets and tests the hypothesis H_0: $\mu_1 = \mu_2$ (or H_0: $\mu_1 - \mu_2 = 0$) against H_1: $\mu 1 \neq \mu_2$.

Example 3:

A third example for a statistical test procedure is a dispersion test for colony counts on parallel plates. In case these plates were incubated from a well mixed suspension with no disturbing influences the counts should be Poisson distributed. The steps to test this are

- formulate the hypothesis H_0: the counts follow a Poisson distribution with the same value for mean and variance
- decide on a sample size, for instance a number of plates of $n = 6$
- decide on the significance level α, for instance $\alpha \leq 0.05$
- choose the test statistic to be calculated from the $n = 6$ results; for this test problem the dispersion statistic D^2 is appropriate. In case the counts are in fact Poisson distributed D^2 follows a χ^2-distribution with $(n - 1) = 5$ degrees of freedom. D^2 will exceed the critical value $\chi^2_{n-1;\alpha} = 11.07$ only with a probability of 0.05 in case H_0 is true. Therefore, the decision rule is to reject H_0 if $D^2 > 11.07$, otherwise not to reject H_0.

The same kind of test could be performed using the log-likelihood ratio G^2 as test statistic instead of D^2.

For the example data set in Section 2, the values for both D^2 and G^2 are below the critical limit, hence there is no evidence to reject H_0.

Note: The decision rule is always formulated in a manner which controls that the significance level α will be met. This means that if the hypothesis H_0 *really is true*, one will get a test statistic beyond the critical limits only with a probability of α, or, in other words, *one will reject H_0 falsely* only in 5% of the cases.

If a false-rejecting rate of 5% is regarded as too often, a smaller α value should be chosen giving different critical limits.

On the other hand, it has to be considered that rejecting H_0 falsely is not the only error that can occur. An error of a second kind would be not to reject H_0 in case it is *not true*.

	H_0 is true	H_1 is true
reject H_0	type 1 error prob. α	right decision prob. $(1-\beta)$
do not reject H_0	right decision prob. $(1-\alpha)$	type 2 error prob. β

However, it is much more difficult to control the probability β of this type 2 error as the alternative hypothesis is usually not as specific as the test hypothesis so that it is difficult to derive expected values of the test statistic if H_0 does not hold.

In fact, whereas the value for α is fixed within a test procedure, there can be an indefinite number of β values depending on how global and indefinite the alternative hypothesis is formulated. Therefore, one often refers to the *power-function* of a test procedure which is defined as the function giving the values $(1 - \beta)$ for all situations summarized in the alternative hypothesis H_1. The power of a test can be interpreted as the probability to detect deviation from the test hypothesis H_0; the values $(1 - \beta)$ should become larger the more different reality and H_0 are.

Thus it is possible to compare different test procedures — testing the same hypothesis on the same significance level but using different test statistics and critical values — by comparing their power for a specified relevant deviation from H_0.

5.   Analysis of variance

The idea behind the statistical techniques called *analysis of variance* (ANOVA) is to compare different groups of data in order to check whether they have the same *location* or not, though this is not what is suggested by its name. The main characteristic of the groups is that they are defined by categories of an influencing factor which is to be investigated.

For instance, if the influence of different nutrient media for colony counts should be investigated, one could take suspensions of sample material and inoculate a number of plates for each media and compare the counting results for these groups (see also Chapter 7; here only analyses of variance for the special problem mentioned in Chapter 7 will be introduced).

However, a main assumption of the ANOVA techniques introduced here is that they assume normally distributed measurement values for all groups having the same variances. For Poisson distributed values — like colony counts should be — this is problematic because if they show differences in location there will be differences in variability as well as the equality of mean and variance which is a main characteristic of this model. In this case, taking square roots of the data would give approximately normally distributed values with a standardized variance of 0.25.

Example 1. One-way ANOVA:

Assume the only factor to be investigated is the effect of $m = 3$ different nutrient media M1, M2, and M3 on the number of colonies that will be counted. From one suspension $l = 5$ plates were inoculated for each medium, yielding 3 groups of counting results:

M1	M2	M3
18	32	45
16	35	47
14	26	52
31	38	68
25	45	56

These counting results are square-root transformed to standardize their variances. This gives the following data set X:

M1	M2	M3
4,24	5.66	6.71
4.0	5.92	6.86
3.74	5.1	7.21
5.57	6.16	8.25
5.0	6.71	7.48

The overall mean of transformed values is $\bar{x}_{..} = 5.907$, the overall variance is $s_{..}^2 = 1.763$.

If there is no influence of the media on these values, the overall mean $\bar{x}_{..}$ and the group means $\bar{x}_{.1}$ for M1, $\bar{x}_{.2}$ for M2, and $\bar{x}_{.3}$ for M3 should be quite similar as well as the overall variances $s_{..}^2$ should exceed the within-group variances no more than could be explained by random variation of the group means alone.

A check for location differences can be performed by comparing the variance of group means $\bar{x}_{.1}, \bar{x}_{.2}, \bar{x}_{.3}$ with the replication variance within groups. If the variance of group means significantly exceeds the replication variance, the hypothesis of equal group means (or of same location) can be rejected thus pointing to systematic differences between the nutrient media.

The values in the example give group means of $\bar{x}_{.1} = 4.51$, $\bar{x}_{.2} = 5.91$, $\bar{x}_{.3} = 7.302$ and group variances of $s_{.1}^2 = 0.572$, $s_{.2}^2 = 0.355$, $s_{.3}^2 = 0.371$.

The overall variance $s_{..}^2$ can be regarded as consisting of a so-called *sum of squares* (ss), which is divided by the *degrees of freedom* (df):

$$s_{..}^2 = \frac{1}{df} \cdot ss = \frac{1}{n \cdot m - 1} \sum_{i=1}^{l} \sum_{j=1}^{m} (x_{ij} - \bar{x}_{..})^2$$

This overall sum of squares can be separated into two parts, one accounting for the deviations of group means from the overall mean, the other accounting for the replication deviations.

$$\sum_i \sum_j (x_{ij} - \overline{x}_{..})^2 = \sum_i \sum_j (\overline{x}_{.j} - \overline{x}_{..})^2 + \sum_i \sum_j (x_{ij} - \overline{x}_{.j})^2$$

or

ss(total) = ss (media) + ss(replications)

with a corresponding split of degrees of freedom

$(m \cdot l - 1) = (m - 1) + (m \cdot l - m)$

or

df(total) = df(media) + df(replications).

For the example data set, these sums of squares are listed in Table 1 together with their degrees of freedom and the resulting *mean sums of squares* (ms) which are derived by dividing the sums of squares by the appropriate degrees of freedom.

The ratio ms(media)/ms(replications) is a so-called F-ratio which follows an F-distribution with $(m - 1)$ and $(n \cdot m - m)$ degrees of freedom in case there are no differences between groups. (F-values are tabulated in statistical tables.)

In the last column the corresponding p-value for the calculated F-ratio is given. This p-value states the probability to get the calculated F-ratio (or an even higher one) in case the hypothesis is true. If this p-value is smaller than the chosen significance level, the hypothesis can be rejected.

Table 1

Analysis of variance table for one-way ANOVA

	df	ss	ms	F	p
Media	2	19.48821	9.74411	22.50596	0.0001
Replications	12	5.19548	0.43296		
Total	14	24.68369	1.76312		

A table such as Table 1 is a common way to summarize the results of an analysis of variance.

Example 2. Two-Way ANOVA:

The example given in Chapter 7 is somewhat more complicated as there is not only one influencing factor but two: different types of nutrient media M1, M2, M3, and different suspensions S1, S2, S3, S4, S5 with which the plates were inoculated. This gives a set of $n \cdot m = 5$ · 3 different combinations of factor categories which allow us to investigate the influence of both effects.

As there might be some interaction between the two factors, $l=2$ replications were made for each combination. This gives some information on the within-combination variance, allowing us to distinguish an interaction effect from this.

The table of square-root transformed values for the data in Table 7.1 is given below:

	M1	M2	M3
S1	3.32	5.66	6.71
	4.0	4.47	5.74
S2	3.61	5.92	5.2
	2.45	6.24	6.16
S3	3.74	5.1	7.21
	4.24	5.48	7.35
S4	7.14	7.42	10.0
	7.28	7.35	9.8
S5	5.0	8.06	9.22
	4.8	9.17	7.55

This data set gives an overall mean of $\bar{x}_{...} = 6.180$ and an overall variance of $s^2_{...} = 3.848$.

The means for the three nutrient media are $\bar{x}_{.1.} = 4.558$, $\bar{x}_{.2.} = 6.847$, $\bar{x}_{.3.} = 7.494$.

The means for the five suspensions are $\bar{x}_{1..} = 4.983$, $\bar{x}_{2..} = 4.93$, $\bar{x}_{3..} = 5.52$, $\bar{x}_{4..} = 8.165$, $\bar{x}_{5..} = 7.3$.

The means for the factor combinations $\bar{x}_{ij}$ are given in the following table:

	M1	M2	M3
S1	3.66	5.065	6.225
S2	3.03	6.08	5.68
S3	3.99	5.29	7.28
S4	7.21	7.385	9.9
S5	4.9	8.615	8.385

The sum of squares contained in the overall variance can now be split into the parts:

ss(total) = ss(media) + ss(suspensions) + ss(interaction)
            + ss(replication)

or as a formula:

$$\sum_{i=1}^{n} \sum_{j=1}^{m} \sum_{k=1}^{l} (x_{ijk} - \bar{x}_{...})^2$$

$$= \sum_i \sum_j \sum_k (\bar{x}_{i..} - \bar{x}_{...})^2 + \sum_i \sum_j \sum_k (\bar{x}_{.j.} - \bar{x}_{...})^2$$

$$+ \sum_i \sum_j \sum_k (\bar{x}_{jk.} - \bar{x}_{i..} - \bar{x}_{.j.} + \bar{x}_{...})^2$$

$$+ \sum_i \sum_j \sum_k (x_{ijk} - \bar{x}_{ij.})^2$$

with degrees of freedom of

df(total) = df(media) + df(suspensions) + df(interaction)
            + df(replications)

or

$$n \cdot m \cdot l - 1 = (m - 1) + (n - 1) + (m - 1) \cdot (n - 1) + m \cdot n(l - 1)$$

The values for degrees of freedom, the sums of squares, and the mean sums of squares for the example data are given in Table 2, together with the F-ratios gained as
- ms(interaction)/ms(replications)
- ms(suspensions)/ms(interaction)
- ms(media)/ms(interaction)

The last column again shows the corresponding p-values for the calculated F-ratios in case in reality the hypothesis of no factor or interaction effect is true.

Which are the appropriate F-ratios to be calculated to perform a statistical test very much depends on the experimental design that forms the background to the data. This itself, however, very much depends on the microbiological problem that should be tackled. What is shown here is only one possible technique to get the mere calculation results for analyses of variance with one or two factors with fixed effects.

The choice of appropriate F-ratios to look at and their interpretation can only be discussed in collaboration with microbiologists and statisticians. Such a discussion should deal with the type of effects that have to be analysed, whether they are random or fixed, with the microbiological meaning of replication variance and interaction, whether there are only main effects with interactions or a hierarchical structure, whether the problem is one of qualitative significance

Table 2

Analysis of variance table for two-way-ANOVA

	df	ss	ms	F	p
Media	2	44.51729	22.25864	17.04183	0.0013
Suspensions	4	51.74841	12.93710	9.90500	0.0034
Interactions	8	10.44895	1.30612	4.02771	0.0098
Replications	15	4.86425	0.32428		
Total	29	111.57890			

testing or one of quantitative estimation of variance components, and so on.

For many different problem formulations, calculation techniques are described in the literature and contained in statistical programs. The challenge is to decide which model and experimental design suits the microbiological problem at hand.

6. References

Campbell, M.J. and Machin, D., 1990. Medical Statistics. A Commonsense Approach. Wiley.

Jarvis, B., 1989. Statistical Aspects of the Microbiological Analysis of Foods. Elsevier, Amsterdam.

Miller, Jr., R.G., 1986. Beyond ANOVA, Basics of Applied Statistics. Wiley.

Documenta Geigy, 1962. Scientific Tables. Geigy (UK) Ltd. Manchester.

Fisher, R.A. and Yates, F., 1974. Statistical Tables for Biological, Agricultural and Medical Research. Longman Group, London.

Heisterkamp, S.H., Hoekstra, J.A., van Strijp-Lockefeer, N.G.W.M., Havelaar, A.H., Mooijman, K.A., in't Veld, P.H., Notermans, S.H.W., Maier, E.A. and Griepink B., 1993. Statistical analysis of certification trials for microbiological reference materials. Report EUR 15008 EN, Office for Official Publications of EC, Luxembourg.

Appendix A

An example of a membrane filter test, with statistical evaluation, is given overleaf.

I	General							
	Laboratory					RIVM-WL		
	Test done by					MB		
	Date					2/4/93		

II	Date of filter batches	Filter batch 1 (old)	Filter batch 2 (new)
	Manufacturer	Sartorius	Sartorius
	Catalog number	13906 47 ACR	13906 50 ACR
	Batch number	1190 13906 9005591	1292 13906 9202173
	Batch size	10000	10000
	Date of purchase	02-92	03-93

III	Visual test of membrane filters	Filter batch 1 (old)	Filter batch 2 (new)
	Number of filters with hydrophobic areas	0	0
	Number of filters with wrinkles	0	0
	Number of brittle filters	0	0
	Test result	approved	approved

V	Colony counts per filter		Filter batch 1 (old)	Filter batch 2 (new)
		1	72	49
		2	83	63
		3	80	64
		4	63	56
		5	74	83
		6	63	75
		7	81	77
		8	60	68
		9	64	60
		10	74	62
		11	74	56
		12	93	75
		13	84	64
		14	93	71
		15	89	62
		16	91	72
		17	69	78
		18	71	71
		19	93	70
		20	71	65
		Batch sum	1542	1341
		Sum squares	121144	91289
		# observations	20	20
		Mean	77	67
		T1 per batch	29.3	20.5

Critical values for T1 (95%)

df	2.5% (<)	97.5% (≥)
16	6.91	28.8
17	7.56	30.2
18	8.23	31.5
19	8.91	32.9

VI	Was the test well performed?	
	Repeat the test if more than three counts are missing	no
	Repeat the test is one or both means are <50 or >80	no
	Repeat the test once if T1 is greater than the critical value	no
	Reject the batch if T1 is greater than the critical value for a second time	no

VII	Significant difference in recovery?	
	T2	14.01
	Reject filter batch 2 (new) if the mean is less than that of filter batch 1 (old) and if T2 > 6.63	Rejected

Appendix B

Preparation of test sample for first-level quality control of plate and membrane filter counts

- Culture a suitable test strain, which has previously been checked for purity and characteristic reactions on the (selective) media, in a non-selective broth under optimal conditions for that strain.

- Dilute in sterile skimmed milk (not rehydrated skimmed milk powder) to 10^{-8}.

- Plate out dilutions 10^{-5} to 10^{-8} on a suitable non-selective medium (spread plates) and on the selective medium to be checked (using the routine inoculation method), and incubate under appropriate conditions.

- Store all dilutions at $5 \pm 3°C$.

- Count all plates, check non-selective plates for purity and selective plates for characteristic reactions. Compare plate counts and the ratio of counts on selective and non-selective media with previous experience. Accept or reject the culture.

- If the culture is accepted, prepare a sufficient volume of test suspension from the dilution series kept in the refrigerator: measure the desired volume of skimmed milk in a dispenser flask, cool on melting ice and add a sufficient volume of a suitable dilution to produce the desired count (for QC of spread-plates by positive controls it is advisable to aim at a final concentration of colony forming units of ca. 100 per 0.1 ml, for pour-plates of ca. 100 per 1 ml and for membrane filtration of ca. 50 per 1 ml; for negative controls and target controls a target concentration of 1000–5000 per 1 ml is advisable).

– Distribute the suspension in volumes of ca. 1.1 ml over cryo-tubes or small, capped propylene centrifuge tubes.

– Keep 10 vials apart for plating on the selective medium.

– Freeze all other vials rapidly by immersing for at least two minutes in a 96% ethanol dry ice mixture prepared by adding dry ice to ethanol in a solid (e.g. stainless steel) vessel until vigorous gas development has terminated; dry ice must be present at all times; alternatively, use liquid nitrogen.

– Take the vials from the liquid, place on boxes and store at –70°C in a biofreezer.

– The following day, thaw 10 vials by immersing in a 37°C water bath, and examine as above.

– Count the plates, check for randomness (using the T_1 statistic described in Chapter 8, Section 2.3(c)) and for survival during freezing; compare with limits set on the basis of experience with test batches and accept or reject the batch of QC samples.

– Check the stability by thawing and plating samples at regular time intervals; for all data, check the T_1 parameter and analyse all data from all time points by linear regression on the $\log_{10}$-transformed counts. If a stable count has been achieved (i.e. if the slope of the regression line does not differ significantly from zero) accept the batch for QC purposes.

Index